BI

VIETNAM,
CAMBODIA and
LAOS

Peter Davidson

BLOOMSBURY
LONDON · OXFORD · NEW YORK · NEW DELHI · SYDNEY

POCKET PHOTO GUIDE

Bloomsbury Natural History
An imprint of Bloomsbury Publishing Plc

50 Bedford Square
London
WC1B 3DP
UK

1385 Broadway
New York
NY 10018
USA

www.bloomsbury.com

First published by New Holland UK Ltd, 2009 as
A Photographic Guide to the Birds of Vietnam, Cambodia and Laos
This edition first published by Bloomsbury, 2016

British Library Cataloguing-in-Publication Data
A catalogue record for this book is available from the British Library.

Library of Congress Cataloguing-in-Publication data has been applied for.

ISBN: PB: 978-1-4729-3284-6
ePDF: 978-1-4729-3282-2
ePub: 978-1-4729-3283-9

2 4 6 8 10 9 7 5 3 1

Designed and typeset in UK by Susan McIntyre
Printed in China

To find out more about our authors and books visit www.bloomsbury.com.
Here you will find extracts, author interviews, details of forthcoming events
and the option to sign up for our newsletters.

CONTENTS

INTRODUCTION

This guide broadly introduces the birds, birding opportunities and related conservation issues in Vietnam, Cambodia and Laos (officially Lao PDR). It is designed for both novices and experienced birders, providing notes on bird status and distribution, identification, vocalizations, behaviour, ecology and conservation. Well over 820 bird species have been recorded in the region. This guide cannot cover all of them, so it focuses on the species most likely to be encountered on even the shortest of visits, as well as a selection of the region's specialities (the endemic and rare birds). Photographs of many of this latter group are published here, in printed form, for the first time.

ORNITHOLOGY IN THE REGION

Vietnam, Cambodia and Laos lie at the convergence of the Himalayan, Chinese and Sundaic biological realms, and many species have evolved here that occur nowhere else on the planet. Most of our knowledge of the region's birds comes from two periods, the early 20th-century colonial era, when it was known as (French) Indochina, and the last two decades. The most comprehensive work on the region's birds is Craig Robson's *Field Guide to the Birds of South-East Asia* (2000, second edition 2009). Anyone with more than a passing interest in the region's birds should have 'Robson' as a field companion.

Birds are ingrained in the culture and traditions of many peoples of the region, but there are few native birdwatchers in any of the countries. Many birders visit the region from elsewhere. The Oriental Bird Club (www.orientalbirdclub.org) is the largest forum and focal point for all matters related to birds and their conservation in this region and Asia as a whole. We still have a great deal to learn about birds in Vietnam, Cambodia and Laos, and amateur ornithologists have a crucial role to play in filling in these gaps in knowledge.

THE ENVIRONMENT AND BIRD HABITATS

A simplified overview of the geography and climate provides some insight into the region's biological richness and uniqueness. Forests once covered most of the land surface, with evergreen formations dominating in moister areas, and deciduous forests in drier areas. Forests remain the dominant land cover today, and the majority of the region's birds are associated with forests.

The mountainous north (rising to about 3,000m) can be thought of as an extension of the Himalayas. To the south, the mountains become sandwiched between the Mekong plains to the west and Vietnam's narrow coastal plain to the east, forming a mountainous spine that extends along the entire length of the Laos–Vietnam border, known as the Annamite Mountains. The natural vegetation of these mountains is tropical evergreen broadleaved forest, mixed with some deciduous and pine components in drier areas on poorer soils. Two seams of forested karst limestone transect the northern and central Annamites.

There are three evergreen-forested massifs isolated from the Annamite range: the Da Lat and Di Linh plateaux in southern Vietnam, the Cardamon and Elephant Mountains in south-west Cambodia, and the Bolaven Plateau in southern Laos. Each supports species and subspecies of bird, other animals and plants that are found nowhere else.

The Mekong River, its tributaries, associated lowland plains and delta dominate western Laos, much of Cambodia and the southern tip of Vietnam. The river-channel sand and rock bars support a distinctive and increasingly threatened bird community. During the May–October rainy season, the Mekong swells and floods huge areas of the adjacent plains in Cambodia and southern Vietnam. The river has an amazing natural flood-regulation mechanism: at the onset of the rains in May, water flows from the Mekong up the Tonle Sap River and into the Tonle Sap Lake, South-East Asia's largest freshwater body, so for six months it flows in one direction, and for the other six months in the opposite direction. The Tonle Sap floodplain and Mekong Delta support extensive grassland and traditional deepwater-rice ecosystems that have their own unique, globally threatened bird communities.

The lowland plains of southern Laos, and northern and eastern Cambodia, are naturally vegetated with an open woodland savannah peppered with small seasonal wetlands, referred to in this guide as deciduous dipterocarp forest. This is the most extensive remnant tract of an ecosystem that once stretched across much of Thailand and Myanmar, and supported the highest density of large mammals outside the East African plains. You will be lucky to see any large mammals there today, but it still supports a highly distinctive bird fauna that benefits greatly from man-made wetlands created during the Angkorian era, between the 9th and 14th centuries.

The Mekong River, and Red River in north-east Vietnam, both disgorge into the South China Sea in large deltas, creating extensive estuaries, intertidal flats and mangroves. Other coastal habitats important for birds include sandy beaches and offshore islands along the South China Sea and Gulf of Thailand coasts.

The region's climate is strongly influenced by the south-west monsoon, which delivers rain to southern Vietnam, Cambodia and Laos between May and October. Towards the end of the intervening November to April dry season, the grassland and open woodland landscapes of the lowland plains become uncomfortably hot. In central and northern Vietnam, the influence of the north-east monsoon bringing moisture off the South China Sea between October and April gives rise to a different climatic regime, with some eastern slopes and low passes in the central Annamites experiencing 'ever-wet' conditions, and some parts of northern Vietnam a distinctly chilly winter.

CONSERVATION

The three countries share similar biological characteristics, but differ enormously in their culture and demography, and are each at quite different stages of emergence from a shared recent era of civil and international conflict. This poses many challenges to development, and results in conservation being a difficult long-term priority to balance alongside the overwhelming immediate issues of poverty, food security, health and infrastructure development that each nation faces. Tourism is becoming an increasingly important component of each country's economy and political decision-making process. As a bird-watching expatriate or tourist, you are playing a role in raising awareness of the importance of wildlife and conservation to the economy.

The region's major bird-conservation issues are largely generic to the planet, with one notable exception: hunting. Habitat loss and degradation driven by urban and rural development, commercial logging and agricultural intensification are reducing forest, wetland and grassland cover. Intact lowland forest (especially on the plains) has disappeared from very large parts of the region. Widespread plantation initiatives threaten much that remains, so that even relatively remote parts of Laos and Cambodia may have lost all extensive lowland forest outside (and possibly within) protected areas within a decade or so. Mature montane forest is generally more extensive but faces similar threats. Development of aquaculture (fish and shrimp ponds) is removing mangrove forest and reclaiming land from major estuaries and mudflats, and from the few large wetlands that remain on the densely settled Mekong plain of Laos and Cambodia. Increasing national and international demands for energy are driving many hydroelectric dam initiatives on most of the region's major rivers, threatening riverine and lowland forest communities. One issue that stands out from many other parts of the world is hunting. Across most of the region, hunting wildlife, including birds' eggs, chicks and adult birds, for subsistence, the cage-bird trade and some medicinal uses is mainstream, steeped in centuries of tradition, and has locally extirpated species or suppressed bird densities.

Each country has a protected areas system and one or more government agencies with mandates to protect wildlife and habitats, but the reality is that they face an overwhelming task with minimal incentives, capacity and resources. Several international conservation organizations provide crucial support at national and local levels. They include BirdLife International (www.birdlifeindochina.org), the Worldwide Fund for Nature (www.panda.org), the Wildlife Conservation Society (www.wcs/international/asia), TRAFFIC (www.traffic.org), Conservation International (www.conservation.org) and Fauna and Flora International (www.fauna-flora.org/asiapacific.php). Several of these organizations are open to volunteer help, so if you plan on staying in any of the countries for a while, do consider this – it could get you into some marvellous birding areas in return for providing much-needed project support.

USING THIS GUIDE AND IDENTIFYING BIRDS

Each species described in this book is illustrated with at least one photograph. It will not always be possible to identify every species from the photograph(s). If the male and female, breeding and non-breeding, or adult and juvenile plumages differ, and the bird you are looking at does not match the photograph, the text covers the important points to help you arrive at a positive identification.

Taxonomy, Species Names and Sizes

The standard reference work for the region is Robson's (2000) *Field Guide to Birds of South-East Asia*. This guide follows the taxonomy and names used by Robson, with amendments to babblers and laughing-thrushes made subsequently by Collar and Robson (2007) in the *Handbook of the Birds of the World*. The length, or variation in length, in centimetres, of the bird from bill-tip to tail-tip is taken from Robson.

Status, Distribution and Abundance

Each species account generally begins with information on the bird's distribution, seasonal status (for example resident, winter visitor, passage migrant), and abundance. This information helps to narrow down the probabilities when trying to identify a bird.

Resident birds (those that remain in the same area all the year round) include partial migrants that move very short distances, for example up and down an elevation gradient between summer and winter. Longer distance migrants fall into the categories winter visitor, breeding visitor and passage migrant.

Winter visitor refers to a species that visits the region only during the northern winter.

Breeding visitor refers to the (relatively few) species that come to the region only to breed.

Passage migrant indicates that the species migrates through the region during the northern spring or autumn, but neither breeds nor over-winters there.

Conservation Status

Eighty-six species (approximately 10 per cent of the total) recorded in the region are currently globally threatened (46) or near threatened (40) with extinction, a proportion typical of many tropical regions. BirdLife International (www.birdlife.org), a global partnership of over 100 conservation organizations, assesses which species are of global conservation concern, placing each in a category from Critically Endangered (facing a very high risk of extinction in the wild) through Endangered and Vulnerable to Near Threatened (facing a relatively high risk of extinction in the wild), based on standard criteria (www.iucnredlist.org/info/categories_criteria). A selection of these species is included in this guide, with notes on the threat(s) each faces.

Habitats

Many species prefer a particular habitat or habitats. Recognizing these habitats is a useful identification tool. Habitats are referred to in this guide in the broadest terms, most of which are self-explanatory, except perhaps for forest types.

Evergreen forest refers only to broadleaved evergreen forest, and not to coniferous forest (which is chiefly pine in this region). Evergreen forest is highly variable, dry in some areas and extremely wet in others, but usually rather dense, with multiple layers from understorey to canopy. Lowland evergreen forest (below 800m) can be very tall and is dominated by trees of the dipterocarp family, while montane evergreen forest (above 900–1,000m) tends to be lower in stature.

Mixed deciduous forest contains a higher proportion of deciduous than evergreen, is multi-layered, often with bamboo in the understorey, and is most obvious during the dry season (November–April), when many deciduous trees drop their leaves due to water stress.

Deciduous (or dry) dipterocarp forest is dominated by just four or five large-leaved dipterocarp tree species, and has an open, woodland-like appearance, with a grassy and herbaceous ground layer; it is often interspersed with patches of savannah and seasonally flooded.

Secondary growth refers to any forest type that is regenerating after being cut or burned.

Altitude

Altitude is another useful aid to identification, because many similar-looking species occupy different altitudinal bands. In this guide, lowland refers to areas below about 800m; mountains (or montane) refers to areas above about 1,000m; and the terms hills and mid-elevations refer to altitudes of about 600–1,500m.

Identification – Plumage

The text describes features and plumages not illustrated in the photograph(s), and discusses how to distinguish similar species. Many of the birds are fairly straightforward to identify with a reasonably good view. Hone your identification skills by taking notes. Aspects to focus on are:

The size of the bird (for example, 'sparrow-sized', or 'a little larger than a pigeon'), but be cautious about judging the size of a solitary bird with no nearby comparator.

The shape of the bird (for example, tall and long-necked, plump and short-tailed), and the shape of the bill and length of the legs.

'Jizz' – a term derived from the phrase 'general impression, size and shape', which refers to a combination of characters that give a bird a distinctive 'feel'; it is something that becomes more familiar with time.

The colour of, and markings on, the different feather tracts. This is absolutely essential to making a correct identification, and learning the different feather tracts, or parts of the plumage, will help you enormously in becoming a proficient birder.

It is only possible to summarise the most important identification features in this guide, so purchase a copy of Robson for truly comprehensive identification information, including key features for distinguishing subspecies, and more detail on distribution, global range and breeding biology.

Identification – Voice

Birds' songs and calls are described when they will assist with identification, or are simply a prominent or interesting trait of the species. In forests you may hear many more birds than you see, so 'birding by ear' becomes particularly useful. Listening to recordings is very worthwhile. One of the most comprehensive libraries of recordings relevant to Vietnam, Cambodia and Laos can be browsed at www.shortwing.co.uk However, there is no substitute for learning the vocalizations in the field. This can seem daunting at first, but set yourself manageable goals, one sound at a time, and you will find that trying to track down a particular sound to the bird delivering it is both a very rewarding experience, and an excellent way of committing songs and calls to memory.

Ecology and Behaviour

Many species, or species groups, behave in distinctive ways. The more time you spend simply watching birds doing what they do, the more familiar you will become with their behavioural quirks, how they feed and what they feed on, how they fly, and so on. This not only assists identification, but also tells you a lot about a bird's ecology – for example, whether it is gregarious or solitary, a frugivore or an insectivore.

Some species behave differently in different areas. If you travel extensively in Vietnam, Cambodia and Laos, you will start to notice that some species are shy (due to local hunting pressure) in certain areas, and much more approachable in others (where they are not hunted as much). The region's strong seasonality dictates key aspects of bird ecology – for example, most insectivorous birds breed between the mid-dry and early wet seasons (January–June, with birdsong peaking during this period), and most water birds breed between the mid-wet and mid-dry seasons (September–February). Many birds form mixed-species flocks – known as bird waves – outside the breeding season.

TIPS FOR FINDING BIRDS

Binoculars will enhance your experience enormously and greatly facilitate identification. Most birders choose 8×30, 8×40 or 10×40 binoculars (never buy binoculars of magnification >10×). Spend time testing different pairs before making a purchase. Many parts of the region are humid and wet, making moisture-proof and waterproof binoculars an especially valuable investment. Outside forest a telescope can be particularly useful in wetlands, grasslands and open country. Carrying a notebook to scribble notes and sketches in, however rudimentary, should be your next priority. It will enable you to identify more birds than committing what you see to memory for later consultation with a field guide, and it can help you to learn birds more rapidly.

In this region, and the tropics generally, it is best to get out birding early, especially in lowland areas. Bird activity (including birdsong) decreases significantly by mid-late morning, and remains low into late afternoon, so you will reap rewards for an early rise. Scan the skys from the mid- to late morning for raptors and large water birds making use of warm updrafts.

Forest birding is very different from birding in more open habitats. The soundscapes (generated by insects and birds) can be bewildering, and seeing birds through layers of foliage requires a sharp eye and a quick binocular draw. A good place to start is the forest edge, for example along a road or track, where a wide variety of species can be found. Inside forest, dress drably, and if birding in a group, keep conversation to a minimum. Follow tracks and trails, move quietly (avoiding noisy leaf litter) and vary your pace, sometimes moving swiftly to surprise shy birds that might otherwise move away ahead of a slow-moving observer, at other times walking very slowly, stopping periodically and waiting for birds to come through.

In forests, it is not unusual to go for an hour or more without seeing a bird, then all of a sudden to come across an active bird wave. Stay calm and try to get good looks at one or two birds. Make 'pishing' (purse your lips and go 'psshh, psshh, psshh') or squeaking noises to attract flocking birds' attention and engage them for longer; imitating the whistled call of the Collared Owlet is another way of attracting smaller birds. Be on the lookout for fruiting and flowering trees, especially figs, which attract a wide variety of frugivorous, nectivorous and insectivorous species. Listen for rustling sounds on the forest floor that may be ground foragers like pheasants, partridges, pittas or thrushes.

In the rainy season, birds tend to shelter during rain, and become active again during sunny breaks; time your birding accordingly. At the height of the dry season, when water is scarce, waiting near water bodies (for example streams or puddles in forest, or natural and man-made pools in open woodland and grassy meadows) can yield good results, especially during the heat of the day.

WHERE AND WHEN TO WATCH BIRDS

As world birding destinations, these three countries are in their relative infancy. Nonetheless, there are some great known birding spots in each, and a host of locations is being developed specifically for nature tourism that promise to become indelibly marked on the itineraries of birding travel well into the future.

Vietnam has the most advanced tourist infrastructure, and the longest-established, best-known network of birding sites. In Cambodia, recent conservation work has discovered several superb birding areas, which are attracting an increasing number of visitors. Laos is developing a series of sites specifically for ecotourism.

Have your binoculars to hand wherever you are – you could find something of interest almost anywhere you look. Also keep in mind that much remains to be learned about ornithology in the region, so take notes on what you see everywhere.

The best time to visit Cambodia, southern Vietnam and Laos is during the dry season, in December–April, when many species are breeding and at their most vocal. As the dry season progresses, daytime temperatures in the south can get uncomfortably hot, and bird activity greatly reduces outside the early to mid-morning period. Spring (March–May) is the best time to visit the mountainous northern parts of the region, especially northern Vietnam, which experiences cool, wet periods in winter, and hot, humid summers.

Vietnam

In the north, there are several excellent birding sites within striking distance of Hanoi. Good montane forest birding and raptor migration in autumn can be enjoyed at Tam Dao National Park. Cuc Phuong National Park and Phong Nha Nature Reserve have some of the best lowland and limestone forest birding in north-central Vietnam. Xuan Thuy Nature Reserve at the mouth of the Red River is the region's premier location for large numbers of shore birds and coastal water birds. In the far north, birding in the Sapa region will introduce you to the Himalayan elements of the region's bird fauna, and the Tram Ton Pass is an important raptor migration bottleneck in spring and autumn.

In central Vietnam, lowland 'ever-wet' forest at Bach Ma National Park supports a number of regional specialities, and the nearby Lo Xo Pass provides access to mid-elevation forest birding. Moving south, Yok Don National Park offers (limited) tourist access to deciduous dipterocarp forest in Vietnam. The two premier destinations in southern Vietnam, both just a few hours drive from Ho Chi Minh City, are the Da Lat Plateau and Cat Tien National Park. The former supports several endemics in montane evergreen and pine forest, the latter some of the best lowland forest birding in the region.

Detailed information on all these can be found in the Birding in Vietnam section of BirdLife in Indochina's website (www.birdlifeindochina.org). This site also hosts the Important Bird Area directories for Vietnam, Cambodia and Laos, which provide bird and habitat status and conservation information about the region's most important bird conservation areas.

Cambodia

Siem Reap provides the perfect base for exploring several excellent local birding sites, and is the home of the San Veasna Centre for Conservation (www.samveasna.org), which every birder visiting Siem Reap should call in on. Angkor Wat affords a good introduction to lowland forest birding. Two hours north-west of Siem Reap, the Sarus Crane Reserve at Ang Trapeang Thmor, a Khmer Rouge-constructed dam, has excellent wetland, wet ricefield and grassland birding.

Prek Toal, which is accessible by boat from Siem Reap, is a core area of the Tonle Sap Biosphere Reserve, supporting the largest concentrations of breeding large water birds in South-East Asia. Along Route 6 south-east of Siem Reap, excellent grassland birding can be had at Stung and just south-west of Kompong Thom town. To the north, Cambodia's northern plains offer the best deciduous dipterocarp forest birding in the region, accessible through an ecotourism project at the village of Tmatboey (refer to Cambodia Ecotourism section at www.wcs.org, which also includes detailed information on visiting the Tonle Sap sites).

During the dry season, take the boat upriver from Phnom Penh to Kratie and explore the Kampi Pool area just north of the town, for some of the lower Mekong River specialities and Irrawaddy Dolphins. Further east is the Seima Biodiversity Conservation Area, which offers excellent lowland forest bird and mammal watching, as do the deciduous dipterocarp forests of the Srepok Wilderness Area (see www.panda.org).

To the south-west of Phnom Penh is the Phnom Aural Wildlife Sanctuary, with hill and montane evergreen forest birding for the adventurous, and Kirirom National Park, with pine woodland and evergreen forest. Bokor National Park has some excellent plateau evergreen forest birding, and further to the south-west, Sihanoukville (Kompong Som) is a good base for exploring the Cambodian coastline's sandy beaches, and mangrove and estuarine habitats.

Laos

Good introductory birding can be enjoyed in and around the capital, Vientiane, along the Mekong River's banks, sandbars and adjacent paddies up to 25km upstream and downstream of the city, at Houay Nyang Forest, and at wetlands at Ban Sivilai and Ban Nongpen. Note that in many parts of Laos, bird densities are suppressed due to hunting; they have been recovering near Vientiane, at least, for several years now.

In the far north-west the Nam Ha National Protected Area, and Bokeo Nature Reserve to the south-west, offer good mid-high elevation evergreen forest birding. The forest edge and dry paddy habitats along the road from Luang Namtha to Muang Sing are also worth exploring. This is generally true for anywhere in Laos: many areas have never been checked by birders. In the far north, there is good montane forest and scrub birding at Phou Fa and along the main roads to Phongsaly. In the cultural capital, Luang Prabang, try Kuangxi Waterfalls Forest Park.

Heading east from Vientiane, the Phou Khao Khouay National Protected Area affords some forest birding opportunities, the Pakxan wetlands are worth a look and Phou Hin Phoun National Protected Area, easily accessible from Thakhek, offers good limestone and lowland evergreen forest birding. Heading south from there, there is easy-access wetland birding at Nong Souy in Champhone district, Savannakhet.

In the south, various Mekong River specialities can be found at Khon Falls on the Cambodian border, and birding in roadside forest on the Bolavens Plateau and its eastern flank can be rewarding. The northern part of Xe Pian National Protected Area has an ecotourism project and can be easily visited: many lowland forest species are present. Check out www.ecotourismlaos.com for more ideas.

BIRD TOUR AGENCIES

Several tour agencies operate standard and customized bird tours to Vietnam and Cambodia. Among those with established itineraries and a regional specialization are Bird Tour Asia (www.birdtourasia.com), Birdquest (www.birdquest.co.uk) and Vietnam Birding (www.vietnambirding.com). Tours in Cambodia can also be arranged through the Sam Veasna Centre (www.samveasna.org).

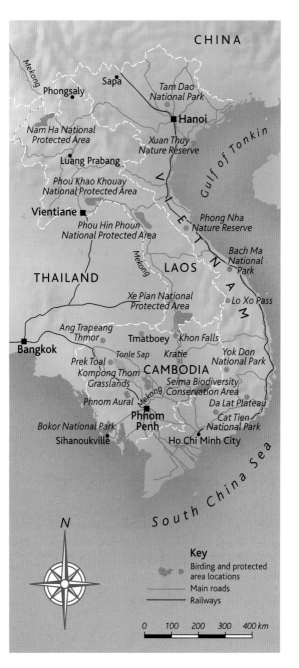

CHINA

Mekong

Phongsaly

Sapa

Tam Dao
National Park

Hanoi

Nam Ha National
Protected Area

Xuan Thuy
Nature Reserve

Gulf of Tonkin

Luang Prabang

Phou Khao Khouay
National Protected Area

Vientiane

V I E T N A M

Phou Hin Phoun
National Protected Area

Phong Nha
Nature Reserve

Bach Ma
National
Park

Mekong

LAOS

THAILAND

Xe Pian National
Protected Area

Lo Xo Pass

Ang Trapeang
Thmor

Tmatboey

Khon Falls

Bangkok

Prek Toal

Tonle Sap

Kratie

Yok Don
National Park

Kompong Thom
Grasslands

CAMBODIA

Seima Biodiversity
Conservation Area

Mekong

Da Lat Plateau

Phnom Aural

Phnom
Penh

Cat Tien
National Park

Bokor National Park

Sihanoukville

Ho Chi Minh City

South China Sea

N

Key

Birding and protected
area locations

Main roads

Railways

0   100   200   300   400 km

11

CHINESE FRANCOLIN *Francolin pintadeanus* 32cm

Dave Farrow

This is a charismatic plump, partridge-like bird that is locally common throughout the region in open forest, grass and scrubby habitats at lower elevations. The male (depicted) has a very distinctive plumage, comprising a mix of bold and intricate black, white and chestnut markings. The female looks similar, only less boldly patterned. The male's territorial call is an emphatic and far-carrying, rather harsh 'wit ta-taak thaa-kow', often delivered from a low branch. This is a familiar sound at dawn and dusk around the fringes of many rural villages and a helpful lead to tracking down this rather shy gamebird, which is resilient to the heavy hunting pressure it faces in many parts of its range.

RUFOUS-THROATED PARTRIDGE
Arborophila rufogularis 26–29cm

Dave Farrow

One of several rather similar-looking, forest-dwelling partridge species in the genus *Arborophila* that occur in the region, this bird inhabits montane evergreen forest in central-northern areas of both Laos and Vietnam. Like other forest partridges, it forages in small groups, shuffling and clearing the leaf litter, and leaving patches of bare earth as it searches for invertebrates, seeds, buds and fruit. It calls with a long monotone whistle that is followed by a series of gradually ascending, whistled couplets, 'who-who, who-who, who-who...'

CHESTNUT-HEADED PARTRIDGE *Arborophila cambodiana* 28cm

James Eaton/Birdtour Asia

This bird is endemic to evergreen forested foothills and mountains of south-west Cambodia. Large-scale logging has reduced and fragmented the habitat within this restricted range, but like other pheasants and partridges this bird is very resilient, largely because it is a prolific breeder. Its plumage resembles that of the Bar-backed Partridge, *A. brunneopectus*, found in similar habitats throughout Vietnam and Laos, but is strongly chestnut on the head and breast (the Bar-backed Partridge has a black and white head with red skin around the eye), with bolder black and white scalloping on the flanks. Look for it along the forest trails on the plateau at Bokor National Park. Its loud territorial calls of see-sawing, whistled couplets, often delivered as a male and female duet, give its presence away.

RED JUNGLEFOWL *Gallus gallus* 41–78cm

Rob Tizard

The wild ancestor of the domestic chicken, this familiar fowl is the most common, widespread, ecologically plastic (found in a wide variety of forest types) and easily seen of the region's pheasants. The males wear gaudy iridescent colours; the female plumage comprises subtle browns and greys. Differentiation from domestic forms can be difficult, and the genetic integrity of wild birds may be threatened by interbreeding with domestic birds ('genetic pollution'). Wild birds have grey legs, and the males have a white rump patch and, in the south only, a white ear patch (as depicted). The species' familiar territorial call is generally slightly higher pitched, with a more abrupt end than that of a farmyard rooster, and is often accompanied by wing-whirring sounds.

SIAMESE FIREBACK *Lophura diardi* 53–80cm

János Oláh

One of at least four *Lophura* pheasants that occur in the region, this one is near endemic, its range extending only into adjacent parts of Thailand. It is uncommon to locally common in lowland evergreen forest types, including forest edge and secondary growth, across much of the region, except northern Vietnam. The male's (depicted) red facial skin, slaty plumage, fire-orange back and long, down-curved, dark greenish tail feathers are unmistakable. The female looks very different, having mostly rufous-brown body plumage with whitish flank scales, and bold blackish and whitish barring on the wings and central tail feathers. Males perform a noisy wing-whirring display, like junglefowl.

GERMAIN'S PEACOCK PHEASANT
Polyplectron germaini 48–60cm

James Eaton/Birdtour Asia

There are two peacock pheasants in the region, with barely overlapping ranges. Germain's Peacock Pheasant is endemic to the evergreen forests of lowlands and hills in southern Vietnam and eastern Cambodia. The feature that characterizes these frequently secretive, brown-plumaged denizens of the forest floor is the multiple white-bordered, iridescent ocelli (or eyes) on their back, wings and tail. Germain's is somewhat smaller and darker than the Grey Peacock Pheasant, *P. bicalcaratum*, and has blood red skin surrounding the eyes. The distinctive call of the male consists of repeated, drawn-out growls or rattles, 'errrraaaaa', leading into a harsher cackling that gets louder and angrier when responding to a rival; the whole series can last several minutes. The bird is common and relatively easy to see at Cat Tien National Park.

GREY PEACOCK PHEASANT *Polyplectron bicalcaratum* 50–76cm

James Eaton/Birdtour Asia

This bird replaces Germain's Peacock Pheasant, *P. germaini*, in the evergreen-forested hills and mountainous areas of northern and central Indochina, north of a line roughly connecting Ratanakiri province in north-eastern Cambodia with the Kontum Plateau in central Vietnam. It is extremely shy and far more often heard than seen. The plumage of the Grey is similar to that of Germain's, but the Grey has flesh-coloured or yellowish facial skin, a more extensive pale throat, a short, bushy crest and narrower pale rings around its ocelli. The Grey has a very different territorial call, a loud, clear and far-carrying '*poor-boy*', the second note inflected upwards, and it also makes a variety of cackling, rattling and growling sounds.

CRESTED ARGUS *Rheinardia ocellata* 74–239cm

William J. Robichaud/NTZWMPA

This extravagantly plumaged pheasant is extremely shy and far more often heard than seen. It inhabits the wetter forest slopes of central-southern Vietnam and central-southern Laos. It is more resilient than its Near Threatened conservation status suggests. Both sexes have blackish-brown body plumage peppered with small white spots. The male (depicted) is unmistakable with its incredulously long (up to 1.75m) and broad tail (the feathers are held vertically). The much smaller female is best told from the peacock pheasants by her bold supercilium and warmer brown body plumage. To see this bird usually requires patience – try 'staking out' a male's dancing ground, a cleared patch of forest floor 60–120cm across, from which it bellows a resonating '*wooa WOW*' call through the forest, usually in February–May.

GREEN PEAFOWL *Pavo muticus* 100–250cm

Allan Michaud

Once very much more widespread and numerous than it is today, a large proportion of this Vulnerable bird's global population is confined to the extensive deciduous dipterocarp woodlands of north-east Cambodia and adjacent southern Vietnam. Both sexes have tall, upright crests, blue and yellow facial discs, glossy green body plumage and caramel flight feathers. The male's 'train', which is over a metre and a half long, is made up of elongated uppertail covert (not tail) feathers. The species is usually shy; check grassy clearings, or try following its loud, disyllabic braying calls ('*Kii-WOW*' and variants) to roost perches at around dawn and dusk. Tmatboey in north Cambodia and Yok Don National Park in Vietnam are good areas to find it.

LESSER WHISTLING-DUCK *Dendrocygna javanica* 38–41cm

Allan Michaud

This is a locally common resident of marshes, lakes and flooded fields across most of the region, but it is absent from parts of northern Vietnam and has much declined in Laos. The longish neck and broad, rounded wings that characterize the whistling-ducks make it readily distinguishable from other duck species in the region. However, the bird's most distinctive feature is its voice, an almost incessantly repeated, high-pitched, whistling '*we-weee*'. The wings also make a whistling sound in flight. The bird is very gregarious, and can form flocks ten-thousand strong when not breeding (Ang Trapeang Thmor in north-west Cambodia is an excellent site to witness this spectacle during the dry season). It tends to feed in fields under the safety of darkness, sticking to larger water bodies by day, and usually nests in tree cavities.

WHITE-WINGED DUCK *Cairina scutulata* 66–81cm

James Eaton/Birdtour Asia

This very large duck inhabits shady forest pools and waterways of lowlands and plateaux. It is Endangered as a result of habitat loss and fragmentation, and hunting, and is now rare everywhere, although birders are finding it with some regularity along the streams of Cambodia's northern plains, near Tmatboey. It is readily recognized by its large size, dark body plumage, mottled whitish head and neck, yellowish or orangey bill and (in flight) bold white patches on the forewing above and below. It generally feeds at night, and flies to and from roost sites in large trees at dawn and dusk, when it often calls with a series of vibrant or harsh honks. Usually encountered singly or in pairs, it nests in tree cavities.

COMB DUCK *Sarkidiornis melanotos* 56–76cm

Rob Tizard

Found in Cambodia and far southern Vietnam, this large duck is a scarce resident of lowland freshwater lakes and marshes. Its blackish upperparts with purplish-blue gloss, contrasting white underparts, and white head and neck peppered with black spots are distinctive. The male is considerably larger than the female (depicted), and has a unique knob (or comb) at the base of his upper mandible, which grows larger during courtship and early breeding (from the start of the wet season). The wholly black wings and white underbody are distinctive in flight. The bird feeds both on land and in water, walks well on land, roosts in trees and nests in tree-holes. Ang Trapeang Thmor and the Tonle Sap floodplain are good areas in which to find it.

SPOT-BILLED DUCK *Anas poecilorhyncha* 55–63cm

James Eaton/Birdtour Asia

This large dabbling duck is quite a common resident of relatively large lowland rivers with rock and sandbars (especially the Mekong), and larger floodplain wetlands, marshes and reservoirs. The feature that most readily distinguishes it from other dabbling ducks is its yellow bill tip. In addition, its brownish, spotted and scalloped plumage is palest on the face and neck, darkest on the rear body and crown, and it has a distinct dark line running through the eye. It has a bold white underwing patch in flight and a white-bordered green speculum. Ducks are hunted in many areas, and are therefore rather wary. This species is usually found in pairs or small flocks, and has a very Mallard-like *'quark'* call.

RUFOUS-BELLIED WOODPECKER
Dendrocopos hyperythrus 19–23cm

This is a startlingly plumaged woodpecker, which is unusual in that it occurs in two distinct populations in the region. One of these is a scarce winter visitor to northern Vietnam (from north-east China), where it occurs in pine, oak and broadleaved evergreen forest; the other is resident in north-eastern Cambodia and adjacent Laos and southern Vietnam, where it inhabits open deciduous dipterocarp woodland of the lowland plains. Some individuals feed by sucking sap from deciduous oak trees. The bird's black and white, ladder-patterned upperparts and bright rufous underparts are unmistakable. The sexes look similar but for the crown, which is red in the male (depicted) and blackish with white speckles in the female.

Pete Morris

GREATER YELLOWNAPE *Picus flavinucha* 32–35cm

Represented by 26 species in the region, woodpeckers are one of the most diverse families. They play a key ecological role as the excavators of tree cavities in which a wide variety of other birds and mammals nest and roost. There are two yellownape species, both characterized by green body plumage and a yellow flare on the nape. The Lesser Yellownape, *P. chlorophus*, has a white cheek bar and barred underparts, and lacks yellow throat sides. In flight, both species have rufous wings, the Greater with bold black bars. Both are common and widespread residents of evergreen and deciduous forests from plains to higher mountains. The Greater Yellownape also inhabits native pine forest. It calls with a loud, far-carrying '*keeaa*' or '*keeape*', often near dusk; the Lesser Yellownape gives a rather plaintive, but also far-carrying '*keeeuu*'.

János Oláh

BLACK-HEADED WOODPECKER *Picus erythropygius* 31–35cm

This has to be one of the most charismatic birds of the dry dipterocarp, deciduous and pine forests of the lowland plains of central-southern Laos, Cambodia and adjacent Vietnam, where it is locally very common. It is often found in vociferous gangs, which utter fabulous, almost maniacal, undulating, yelping laughter. The bird's plumage is equally striking, consisting of a black head (red crown in male only) with pale yellow eyes, yellow underparts (brightest on the breast), and green upperparts with a red rump and dark tail. It also shows small white wing patches as it makes undulating flights between trees.

James Eaton/Birdtour Asia

RED-VENTED BARBET *Megalaima lagrandieri* 30–34cm

Craig Robson

The Red-vented Barbet is endemic to the region and is the largest of the barbets, along with the Great Barbet, *M. virens*, which it replaces south of far northern Laos and far northern Vietnam. It is common in evergreen forests from lowlands up to mid-mountains, particularly in the Annamite Mountains. It is a rather plain barbet with green plumage, a brownish head, an orangey tuft above the base of its very heavy bill and red undertail coverts. It has a throaty, rather ear-jarring, disyllabic territorial call, which is repeated every couple of seconds for minutes at a time and carries across entire hillsides. Like most barbets, it congregates at fruiting trees and spends most time in the canopy, so it can be very hard to see.

GREEN-EARED BARBET *Megalaima faiostricta* 25–27cm

One of two barbets with brown-streaked heads (the other being the Lineated Barbet, *M. lineata*), this is a common resident throughout the region, endemic to Indochina and neighbouring Thailand and south China. It inhabits open evergreen and deciduous forests of lowlands and hills, while the Lineated Barbet is only found in deciduous forest, especially deciduous dipterocarp woodland, at similar elevations, but is absent from central Vietnam. The Lineated has whitish (not green) cheeks and ear coverts, and a yellowish bill and orbital skin around the eye. The Green-eared has a three- or four-note territorial call, the Lineated a strident but mellow, two-note whistle, 'walk-on', with the second note being higher.

János Oláh

GOLDEN-THROATED BARBET *Megalaima franklinii* 21–23cm

János Oláh

This is the most common barbet of montane evergreen forests (above 800m) in Indochina, although locally it occurs in the lower hills in central Vietnam. The bird is found in most areas, except much of Cambodia and far southern Vietnam, and is best distinguished from other barbets of the region by its colourful red, yellow, grey and blackish head pattern. Like other barbets, it frequents the forest canopy, congregates in fruiting trees and nests in tree cavities. It has a repetitive territorial call, a trisyllabic '*tuk-wor-ruk*' that has a rolling, liquid quality. Once one bird starts calling, a whole hillside can erupt in a chorus of Golden-throated Barbets calling antiphonally.

BLACK-BROWED BARBET *Megalaima oorti* 22–23cm

The race of this species found in Indochina is sometimes considered a separate species, the Annam Barbet, *M. annamensis*, and as such is an endemic to the region. It is a fairly common resident of mid-elevation evergreen forests (600–1,450m) within a relatively restricted range that spans eastern-southern Laos, north-east Cambodia and part of central-southern Vietnam (encompassing the Da Lat Plateau). The bird looks somewhat like the Golden-throated Barbet, *M. franklinii*, but has blue head sides and a broad blue lower throat band. It also lacks blue on the wing feathers and has a narrower black eye mask/supercilium. The territorial call of the species consists of a four-syllable, rhythmic '*tok-a-roo-ut*', which is repeated every second or so.

János Oláh

BLUE-EARED BARBET *Megalaima australis* 17–18cm

Pete Morris

A common resident throughout the region, this barbet inhabits both evergreen and mixed deciduous lowland and hill forests, including secondary forest, forest edge and even rural areas with numerous large trees. Its small size, and head pattern consisting of blue throat, ear coverts and crown, orange-yellow cheek patch, red lower cheek and black forehead, separate it from the other barbets. Its calls are unique too, the main territorial call being a rapidly repeated, rather wooden and very monotonous '*t'ruk*', and a series of whistled '*plew*' notes uttered at approximately one per second. Like other barbets, this species congregates in fruiting trees to feed and often sits motionless for long periods, especially when calling.

ORIENTAL PIED HORNBILL *Anthracoceros albirostris* 68–70cm

Rob Hutchison/Birdtour Asia

This is the smallest – and most widespread and numerous – hornbill in the region. It is found in a variety of forest types, including secondary growth, coastal scrub and plantations, from the lowlands up to mid-elevations. The female has more extensive blackish smudging on the bill and casque than the male. Usually found in flocks, which are sometimes quite large and communicate with a vociferous high-pitched, yelping laughter, the bird's distinctive flight action consists of rapid wingbeats interspersed with short glides. Like many hornbills it nests in tree cavities, the entrance to which is partially sealed (to reduce the predation risk), so the female relies on the male or wider social group to bring her food during incubation and early chick-rearing.

GREAT HORNBILL *Buceros bicornis* 119–122cm

Rob Tizard

Now scarce across most of the region and considered Near Threatened globally, this hornbill continues to decline due to hunting, exacerbated by degradation and fragmentation of its preferred low-mid elevation, mature evergreen or mixed deciduous forest habitat. The home range of this massive bird can extend to hundreds of square kilometres. Often the first sign of large hornbills is the sound of air whooshing through their wing feathers, audible at some range. The bird's call is a prolonged series of loud, deep '*kok*' notes, leading into a crescendo of loud, harsh barking. It is generally encountered in pairs or small groups, especially at fruiting trees, and requires large old trees for nesting.

RUFOUS-NECKED HORNBILL *Aceros nipalensis* 117cm

Globally threatened (Vulnerable) due to hunting, and habitat loss and fragmentation, the regional population of this hornbill, which inhabits mature mid-elevation evergreen forests in the Annamite Mountains, along the Lao–Vietnamese border, is probably one of the most important remaining in the world. The male (depicted) is unmistakable. The mostly black female looks like the Wreathed Hornbill, *A. undulatus*, but has white wing tips and distal part of the tail (the tail is all-white in the Wreathed). Both sexes lack casques and have a series of dark ridges at the base of the upper mandible and a bright orange-red gular pouch. The bird's call note, a repeated single '*kup*', is similar in quality to that of the Great Hornbill, *Buceros bicornis*, only softer and does not lead to a barking crescendo.

Rob Tizard

ORANGE-BREASTED TROGON *Harpactes oreskios* 27–31cm

Trogons are long-tailed dwellers of the mid-canopy, with broad, stubby bills that are used to catch invertebrates like cicadas. The Orange-breasted is one of two trogons in the region. It is a widespread resident of evergreen and mixed deciduous forests of lowlands and hills in Laos, Cambodia and central-southern Vietnam. The female is similarly patterned, but duller than the male (depicted). Both sexes have white outer tail feathers, which often flash as the bird flies short distances from one perch to another with an audible whirr of wings. This species is rather shy, usually occurs in pairs and perches very still for long periods on very short legs. Its territorial call is a soft, 3–5-note, whistled 'chew-chew-chew...', delivered in an even tone.

János Oláh

INDIAN ROLLER *Coracias benghalensis* 32–34cm

Rob Tizard

The Indian Roller is a widespread and common resident (except in north-east Vietnam) of open habitats such as agricultural landscapes, deciduous dipterocarp woodlands and urban fringe. Its plumage can appear to be a rather dull purplish-brown at rest, but it bursts into a blaze of colour in flight, when the bright turquoise and brilliant purple-blue feathers on the tail and the wings are revealed. Rollers often rest on prominent open perches like electricity poles, from which they drop down to the ground to catch their prey, including larger insects and reptiles, with their strong blackish bill. Some birds also make aerial sallies in order to catch insects on the wing.

BLUE-EARED KINGFISHER *Alcedo meninting* 16cm

James Eaton/Birdtour Asia

This close relative of the Common Kingfisher, *A. atthis*, is quite a common resident of streams and pools in evergreen and mixed deciduous forests of the lowlands and lower hills in Cambodia, Laos and central-southern Vietnam. It has deeper blue and deeper rufous-orange body plumage than the Common Kingfisher, with (in all but juvenile birds) blue, rather than orange ear coverts. The Common Kingfisher is a frequent winter visitor to rivers and wetlands throughout the region, and a local resident in the north. A careful look is required to separate these two species when encountered along a shady or forested stream. In both, the bill of the male (depicted) is wholly blackish, but the female has a reddish lower mandible. Each species calls with a shrill, abrupt and high-pitched '*si*' or '*pi*'.

WHITE-THROATED KINGFISHER *Halcyon smyrnensis* 28–29cm

This kingfisher is generally a common resident (although locally it has been much reduced by hunting) in open habitats, secondary growth and cultivated areas throughout the region, mostly in lowlands and hills. It is a noisy bird that often occurs some distance from water. Its plumage, in particular the brilliant turquoise upperpart feathers, features in one of the earliest documented uses of animal parts in trade, as decorative adornments for hats and headscarves. Its bold white throat and breast 'bib' (not visible in the photograph) contrast with its chestnut head and underparts. In flight it shows white wing patches. Its call is a frequently repeated staccato laughter, and its territorial call is a loud, whinnying '*kikikikikikikiki*'. It often perches on wires, exposed snags and other prominent locations.

Allan Michaud

25

PIED KINGFISHER *Ceryle rudis* 27–30cm

Rob Tizard

A striking, intricately marked black and white kingfisher, this species is a locally common resident (much reduced in some areas) of larger lowland rivers, canals and lakes in open landscapes, particularly the Mekong River and its floodplain, the Tonle Sap floodplain and the Red River in northern Vietnam. The bird's habit of regularly hovering helps to identify it even as a distant silhouette. When perched, it often raises a short crest. The female (depicted) has one black breast band, while the male has two. This kingfisher is generally found close to earth or sandy banks, in which it excavates its nest-holes. Its call is a series of shrill whistled notes.

COLLARED KINGFISHER *Todiramphus chloris* 24–26cm

Rob Tizard

This kingfisher is a locally common resident of coastal habitats, such as mangroves and tidal creeks in the southern half of the region, and also occurs along the Mekong and Tonle Sap Rivers in Cambodia and southern Vietnam. Its brilliant blue upperparts, variably washed with turquoise, and gleaming white underparts and collar, are unmistakable. Like several of the large kingfishers, it is strongly vocal, and calls with repeated and loud '*kick-you*' shrieks and a shrill, staccato laughing call consisting of a single note repeated. It often perches on overhead wires, and excavates a nest in earth and sand banks, or nests in a tree-hole.

GREEN BEE-EATER *Merops orientalis* 19–20cm

The smallest bee-eater in the region, this is a widespread (absent from northern Vietnam) and common resident of drier open landscapes up to mid-elevations, including deciduous dipterocarp woodland, cultivated areas and coastal dunes. As its name suggests, its plumage is mostly green, with a distinct coppery crown and 'shawl', light blue throat, black breast gorget (which distinguishes it from the similarly plumaged, but brighter and larger Blue-throated Bee-eater, *M. viridis*) and elongated central tail feathers. Like other bee-eaters, it makes regular graceful aerial forays to catch bees and other insects. Its voice consists of a pleasant, quite high-pitched, bubbly trilling.

Allan Michaud

BLUE-TAILED BEE-EATER *Merops philippinus* 23–24cm

This is a locally common and gregarious bee-eater of open lowland landscapes, generally found near fresh water. It is resident in some areas, and a breeding visitor, passage migrant or winter visitor in others. Its body plumage is bright green, with a distinctive head pattern consisting of a black eye-mask, chestnut-orange lower throat and yellow chin, and a bright, pale blue lower back and tail, the latter with elongated central feathers. The bird nests in colonies, excavating nest-holes in riverbanks and other bare ground. Like other *Merops* bee-eaters, it has a coppery-coloured underwing, which is visible in flight. Its call is a rolling, liquid *'prillip'*, regularly repeated, especially in flight. Large roost gatherings occur in some areas, such as the Tonle Sap.

Peter Davidson

BLUE-BEARDED BEE-EATER *Nyctyornis athertoni* 33–37cm

In marked constrast to the *Merops* bee-eaters of more open habitats, the behaviour of this chunky bee-eater is more sluggish, and it lives within the mid-canopy of evergreen and mixed deciduous forests, from lowlands to mid-mountains. Although widespread, it is not numerous, and is generally found singly or in pairs. It has a square-ended tail and mainly dull green plumage, but for a shaggy blue throat and breast feathers, and blue-streaked yellowish lower underparts. Its voice consists of various deep, guttural croaking or harsh cackling sounds. Some individuals will sit motionless for very long periods.

Dave Farrow

LARGE HAWK CUCKOO *Hierococcyx sparverioides* 38–41cm

This member of the cuckoo family superficially resembles the *Accipiter* hawks, having a barred tail and rounded wings, and is colloquially known as the 'brain fever bird' because of its distinctive loud, shrill territorial call, which is often rendered *'brain fee-ver'* (or *'pwee, wee-wur'*). It is a common resident in Cambodia, Laos and northern Vietnam (but more scarce in other parts of that country), and inhabits evergreen forest at all but the highest elevations, generally breeding in hills and mountains. This bird is the largest of three rather similar hawk cuckoos in the region, all of which are quite secretive and hard to observe, and it lays its eggs in (brood parasitizes) the nests of spiderhunters.

Martin Hale

ORIENTAL CUCKOO *Cuculus saturatus* 30–32cm

One of four similar-looking, forest-dwelling cuckoos of the Cuculus genus in the region, this one is fairly common as a breeding visitor to central and northern Laos and Vietnam, and as a passage migrant. The Indian Cuckoo, *C. micropterus*, a fairly common resident in the region, differs from the Oriental Cuckoo in having brownish upperparts and broader dark underpart barring. The Oriental Cuckoo's territorial call in Indochina consists of 2–4 rapid, mellow, even-tone notes, '*wu-wu-wu*'; the Indian's is a four-note 'hu-hu-hu-who', the third note higher, the fourth low. The Lesser and Common Cuckoos, *C. poliocephalus* and *C. canorus*, are very rare outside the far north. The female also occurs in a rufous morph, which looks like a large Banded Bay Cuckoo, *C. sonneratii*, without a supercilium and eye-patch.

James Eaton/Birdtour Asia

BANDED BAY CUCKOO *Cacomantis sonneratii* 23–24cm

Pete Morris

This is a widespread and fairly common resident of evergreen and deciduous forests, including secondary growth, from plains to mid-elevations. Like other cuckoos, it is more likely to be heard than seen, and has a distinctive and easily whistled call, '*we-who, we-who*', delivered rather quickly with emphasis on the first and third notes; it also gives a much longer series of repeated '*we*' and '*pew*' notes on a rising scale. The bird's entire plumage is finely barred, and it is best separated from immature and hepatic (rufous-coloured) forms of other cuckoos (such as the Plaintive Cuckoo, *C. merulinus*) by its whitish supercilium and dark eye-patch. It is usually a solitary bird of the middle and upper canopies.

PLAINTIVE CUCKOO *Cacomantis merulinus* 22–23cm

This is the most abundant and widespread cuckoo of relatively open habitats, found in open woodlands, secondary growth, parks, cultivated areas, gardens and cities, from the plains to the mid-mountains. The male is grey above with peachy-buff underparts, with young males (depicted) showing some rufous barring in the wing and tail feathers. The female sometimes looks like the male, but more usually occurs in a hepatic (rufous-coloured) morph, a trait of several members of the cuckoo family, which is rufous with blackish bars and a whitish supercilium (rather like the Banded Bay Cuckoo, *C. sonneratii*). The Plaintive Cuckoo has a very distinctive territorial call, consisting of a series of loud and plaintive whistles, '*pwee, pwee, pwee, pwee-pwee-pi-pi-pi*', more rapid or slurred and descending in pitch at the end.

Martin Hale

VIOLET CUCKOO *Chrysococcyx xanthorhynchus* 17cm

János Oláh

This beautiful small cuckoo is a widespread but not very numerous resident of lowland deciduous and evergreen forests and also locally, of parks and gardens. Its distinctive sharp '*kee-wick*' call is often the first indication of its presence. The male (depicted) is a stunning glossy violet-purple with orangey bill and eye, and white barring on the underparts. The Asian Emerald Cuckoo, *C. maculatus*, is similarly patterned but glossy green in colour. The female is less glossy, with dull bronzy upperparts and dark-barred whitish underparts, lacking the strong green tones of the similar-looking female Asian Emerald Cuckoo, and has a dark tip to the bill.

CORAL-BILLED GROUND CUCKOO *Carpococcyx renauldi* 69cm

Rob Hutchison/Birdtour Asia

Endemic to Thailand and Indochina, this large, long-tailed and long-legged, ground-dwelling cuckoo is locally common in evergreen forests at low to middle elevations, especially in the Annamite Mountains along the Laos–Vietnamese border, but is shy and hard to see. It is more often heard, repeating its slightly haunting, drawn-out *'whooaaaar'* territorial call every 5–10 seconds; it also gives a more strident, rolling version of a similar sound. Its coral-red bill is striking when seen, and its body plumage is a soft grey, with black head, neck, wings and tail, and a patch of purplish skin around the eye. This cuckoo is not parasitic and generally nests in trees, although it spends most time on the ground.

GREATER COUCAL *Centropus sinensis* 48–52cm

Rob Tizard

Coucals are weak-flying, principally terrestrial birds. Two species are resident in the region, both of which are widespread and common at low-mid elevations. The Greater Coucal is most likely to occur in open forest, forest edge and secondary growth. The Lesser Coucal, *C. bengalensis*, prefers damp scrub and grass. The Greater is larger and heavier billed; the adult (depicted) wears the same plumage throughout the year. In breeding dress, the Lesser looks like the Greater with pale mantle streaks, and in non-breeding plumage it is grey-brown below and heavily streaked above. Both species call with long series of deep and resonant *'hoop'* notes, the Greater's gradually descending then (often) ascending in pitch, the Lesser's ending with a series of higher *'kotook'* notes. Coucals clamber in bushes and up grass stems, and build round nests of grass with side entrances.

VERNAL HANGING PARROT *Loriculus vernalis* 13–15cm

By far the region's tiniest parrot, this is also the area's only representative of a genus whose diversity is centred on the Sundaic and Wallacean regions of Indonesia. Hanging parrots are unique among birds in their apparent ability to sleep upside down. This one is a common and widespread resident of lowland and hill forests, both evergreen and deciduous, although is absent from the northern half of Vietnam. The male has a blue throat, which the female (depicted) lacks. In flight, which is fast and direct, it flashes a turquoise underwing, and calls with a squeaky, high-pitched '*dzee-ip*' or '*p'zee-it*'. It is mainly frugivorous, uses its bill to clamber through tree branches and frequently hangs upside down.

Craig Robson

RED-BREASTED PARAKEET *Psittacula alexandri* 33–37cm

The most common and widespread of the parakeets in the region, this species can still be found in huge flocks in some areas, for example around Angkor Wat and the north-west corner of the Tonle Sap Biosphere Reserve. It inhabits deciduous forests and rural landscapes with relict tall trees, and is readily distinguished from other parakeets by its broad black moustache and reddish-pink breast, and the narrow black bar across its forehead. The male has a bright red bill and the female's bill is black, and their harsh, raucous, screeching calls can be heard from long range. This species is extensively trapped for the cage-bird trade, and is the most common captive parrot from Hanoi to Ho Chi Minh City. It both roosts and nests colonially.

Dave Farrow

ALEXANDRINE PARAKEET *Psittacula eupatria* 50–58cm

Rob Tizard

This is the largest parrot in the region, and also the rarest. Its range and numbers have been greatly reduced as a result of capture for the cage-bird trade – you are more likely to see this bird in a cage in Hanoi or Ho Chi Minh City than in the wild now. That said, the bird is still locally common, its stronghold being the open woodlands of southern Laos, Cambodia and pockets of southern Vietnam. The combination of its large size, very long tail, huge red bill and maroon shoulder patch make it readily identifiable. The male has a distinctive bi-coloured collar, which is black at the front and deep pink around the back. The bird's loud, ringing '*treeooo*' call is distinctive, and it is loosely colonial, nesting in tree-holes excavated by woodpeckers.

HIMALAYAN SWIFTLET *Collocalia brevirostris* 13–14cm

Rob Tizard

Swiftlets are a group of birds whose species-level taxonomy remains poorly understood. This species occurs in forested and open rural landscapes of the foothills and mountains (especially limestone) of northern and central Indochina in both winter and summer. The bird resembles a swift but has a more fluttering flight, paddle-shaped (rather than aerodynamically swept-back) wings and a slight notch in its square-ended tail. Some Himalayan Swiftlets have pale rumps, but these are generally not as pale as those of the very similar Germain's Swiftlet, *C. germani*, which overlaps considerably in range. These extraordinary birds echolocate, and they nest in colonies in caves or on vertical rock faces, constructing their nests from moss and saliva. Their voices consist of a low, rattling twitter.

ASIAN PALM SWIFT *Cypsiurus balasiensis* 11–12cm

This is a very common resident of the region, often found in noisy and active flocks around temple groves, villages and cultivated open country with numerous palm trees – in which it both roosts and nests. Similar in size to swiftlets and entirely dark brown, it is best identified by its very slim, pointed wings and long, slender tail, which is deeply forked when it is (rarely) spread; its shrill, chattering calls are striking. It is much smaller and does not have the white rump of the Crested Treeswift, *Hemiprocne coronata*.

Rob Tizard

CRESTED TREESWIFT *Hemiprocne coronata* 21–23cm

This close relative of the swifts has less aerial habits and a more colourful plumage. It is a fairly common resident of open deciduous forests, forest edge and open evergreen woodlands in the lowlands, foothills and plateaux, but is absent from the northern half of Vietnam. Its long, slender and upright posture when perched, combined with grey body plumage and usually erect crest, are distinctive. The male bird (depicted) has a rufous face; the female has a narrow blackish mask. In flight, the bird has a long, slender tail that is deeply forked when spread, and very long, slim wings, which combined with its size and rather explosive, disyllabic call help to distinguish it from the swifts.

Pete Morris

ORIENTAL BAY OWL *Phodilus badius* 29cm

Smaller than the Barn Owl, *Tyto alba*, which is found in both cities and grasslands throughout the region, this rather enigmatic owl is a widespread but uncommon resident of broadleaved evergreen forests (except in central and north-west Vietnam). It has almost trian- gular facial discs, and calls with a series of eerie, upwardly inflected whistles that have earned it the name '*Pi Con Koi*' in Laos, which loosely translated means 'Forest Spirit', or 'Ghost'. It is strictly nocturnal and a hard bird to see – when tracking a calling individual, look for it perched quite low down, one foot above the other on an upright stem.

James Eaton/Birdtour Asia

ORIENTAL SCOPS OWL *Otus sunia* 19cm

There are three scops owl species in the region, all rather similar in appearance, with short 'ear tufts' (elongate feathers that in actual fact have nothing to do with the ears). All are nocturnal and best distinguished by their territorial calls. The Oriental Scops Owl gives a measured four-note '*toik, toik ta-toik*'; the Collared Scops Owl, *O. bakkamoena*, gives a mellow, falling '*pouu*', repeated every 10–15 seconds, and the Mountain Scops Owl, *O. spilocephalus* (which is not always confined to mountain areas), gives a well-spaced '*poo, poo*', repeated every 5–7 seconds. The Oriental occurs in two morphs, one greyish and the other rufous. Both the Mountain and Collared are found throughout Indo-china, but the Oriental is restricted to Cambodia, southern and central Laos and central Vietnam.

James Eaton/Birdtour Asia

BROWN FISH OWL *Ketupa zeylonensis* 49–54cm

Of the three fish owls in the region (the Brown Fish Owl, and the Buffy and Tawny Fish Owls, *K. ketupu* and *K. flavipes*), the Brown is the most common and widespread. The Buffy occurs locally in Cambodia and southern Vietnam, and the Tawny is a rare resident of Laos and southern Vietnam. All three occur along evergreen-forested streams and rivers at lower elevations, and all look very similar, with prominent floppy ear-tufts and yellow eyes. The Brown has vertical dark underpart streaks with narrow brown crossbars, the Buffy lacks the crossbars, and the Tawny is the largest and has more orangey-buff underparts. Although nocturnal, fish owls can be conspicuous by day, easily flushing from their roost perches in streamside trees.

James Eaton/Birdtour Asia

BROWN WOOD OWL *Strix leptogrammica* 47–53cm

This is one of two wood owls in the region (the Spotted Wood Owl, *S. seloputo*, is the other). Both are uncommon residents, often (although not always) encountered in more openly wooded and treed village settings than other owls. Both have neatly rounded heads and gentle expressions, the Brown with bold dark smudges around the eyes, which the Spotted lacks. The Brown also has dense blackish barring on the underparts. The underpart barring on the Spotted is more spaced and contrasts with the whitish ground colour. The Spotted's typical call note is a loud, abrupt '*Who*', whereas the Brown gives deep, relatively complex strings of '*hoo*' notes running together, and an eerie scream, '*yyyoow*'. Small birds will often mob a roosting wood owl, and indeed other large owls, by day.

James Eaton/Birdtour Asia

COLLARED OWLET *Glacidium brodei* 16cm

This diurnal owl is the smallest owl in the region. It is widespread and not uncommon in evergreen forests at up to 3,000m, and its call often attracts a commotion of smaller birds, although its prey consists primarily of large insects. Imitating the Collared Owlet's four-note 'poop, poo-poop, poop' territorial call (easily whistled) is an old birder's trick to draw in birds that might otherwise remain undetected in the forest's lower storeys. The Collared Owlet is similar to the Asian Barred Owlet, *G. cuculoides*, and best distinguished from it by its small size and the imitation face pattern on the back of its head – probably an adaptation to confuse mobbing birds.

Dave Farrow

ASIAN BARRED OWLET *Glaucidium cuculoides* 21–23cm

This is a common resident throughout the region in deciduous and open-canopy evergreen forests, open country with scattered trees, parks, village gardens and towns (it is now common in central Vientiane). Like the Collared Owlet, *G. brodei*, it is active during the day and has an undulating flight, as well as a distinctive habit of wagging its tail from side to side when agitated. It is larger than the Collared Owlet, with a proportionally larger head and whitish eyebrows, and lacks the imitation face on the back of its head. One of its most regular calls is a long (about ten-second), quivering trill, descending in pitch and increasing in volume. It nests in old woodpecker- or barbet-excavated tree cavities.

János Oláh

HODGSON'S FROGMOUTH *Batrachostomus hodgsoni* 25–28cm

Frogmouths are strictly nocturnal, nightjar-like birds, with similarly cryptic plumage, shorter, rounded wings, very broad bills and a huge frog-like gape for catching insects, and more arboreal habits than nightjars. There are two frogmouth species in the region, Hodgson's Frogmouth, which inhabits mid-elevation evergreen forests of Laos and central Vietnam, and the Javan Frogmouth, *B. javensis*, which dwells in lowland evergreen and mixed deciduous forest in Laos, Cambodia and southern Vietnam. The call of Hodgson's Frogmouth is a series of repeated eerie '*were-iiii*' notes; the Javan gives a three-note whistle, '*tee-loo-ee*', growling '*gwaa*' notes and a maniacal descending laughter.

János Oláh

INDIAN NIGHTJAR *Caprimulgus asiaticus* 23–24cm

Peter Davidson

Nightjars are graceful aerial foragers (of moths, beetles and crickets) with longish slender wings and long tails. They are mainly crepuscular or nocturnal, although their habit of roosting on the ground results in the occasional diurnal encounter, as is obvious from the photograph. All six species in the region have similar cryptic plumages; calls are especially useful for identifying this group. The Indian Nightjar is a locally common resident of open dry woodlands, scrubby regions and even cultivated areas. The male's territorial call sounds like a ping-pong ball bouncing to rest, '*chouk, chouk, chouk-chouk-er-rouk*'. The Indian Nightjar is similar to the more widespread Large-tailed Nightjar, *C. macrurus*, which occurs in deciduous and evergreen forests, and whose territorial call is a loud and resonant '*chonk*' that is repeated at one- or two-second intervals.

SPOTTED DOVE *Streptopelia chinensis* 30–31cm

The Spotted Dove is the near-ubiquitous resident dove of open habitats throughout the region, including deciduous woodlands, forest edge, scrub, cultivated areas, parks, gardens and even dense urban centres. In heavily settled parts of Laos, hunting has eradicated it from some areas. In flight, the bird has a long, wedge-shaped tail with broad white corners, and a pale grey bar across each wing. Be aware of the smaller 'zebra-striped' Peaceful Dove, *Geopelia striata*, a recently introduced colonist on the Mekong plain in Laos and around Siem Reap in Cambodia, and the Oriental Turtle Dove, *S. orientalis*, a widespread but uncommon winter visitor that has rufous-scalloped upperparts, and a grey rump and uppertail. The Spotted Dove's song is a soft, repeated cooing with a very soothing quality, and it is an extremely common bird in the cage-bird trade.

RED COLLARED DOVE *Streptopelia tranquebarica* 23–24cm

This is a rather small, compact dove, which is a common and widespread resident of relatively dry, open habitats with sparse trees, rural settlements, palm groves and deciduous dipterocarp woodland in the lowlands and hills. Adult males have a strongly vinous-pink breast and upperparts contrasting with a grey head (the bird depicted is an immature male). Females are much browner overall. All plumages have blackish flight feathers contrasting with the rest of the upperparts. The song is a repeated soft, rhythmic cooing, '*guru-oo-roo*', soothing like that of the Spotted Dove, *S. chinensis*. This bird, too, is a target for the cage-bird trade, but this threat appears to have little population-level impact. It feeds mainly on the ground, taking grains, grass and herb seeds, as well as buds and young leaves from trees.

THICK-BILLED GREEN PIGEON *Treron curvisrostra* 26–28cm

Dave Farrow

This is the most common and widespread of this distinctive group of pigeons in the region, inhabiting evergreen and mixed deciduous forests up to mid-elevations, away from the coast. It is often encountered in flocks, especially at fruiting trees, and its presence is given away by noisy flapping and jostling for food, as well as its bizarre musical, wavering and softly whistled calls, some beautiful, others almost comical. The plumages of many of the green pigeons are confusingly similar. This species is best distinguished from others by its thick, pale green bill with a red base and blue-green eye-ring. The male (depicted) has maroon upperparts and dull chestnut undertail coverts, the female green upperparts and creamy undertail coverts with dark green scales.

YELLOW-FOOTED GREEN PIGEON *Treron phoenicoptera* 33cm

James Eaton/Birdtour Asia

One of the largest-bodied green pigeons, this is a resident of the deciduous dipterocarp woodlands of the plains of central-southern Laos, central-southern Vietnam and Cambodia. The combination of bright yellow-green breast and neck collar, grey underparts, crown and rear neck, yellow feet and bicoloured tail, which is bright greenish basally, dark grey distally, makes its identification relatively straightforward. It is more often encountered in ones and twos than other green pigeons. The Tmatboey region of Cambodia's northern plains and Yok Don National Park in Vietnam are good places to find it in high abundance.

PIN-TAILED GREEN PIGEON *Treron apicauda* 30–40cm

This is an uncommon resident of evergreen forests and forest edge, in mid-mountains and hills, found in north-east Cambodia, central regions of Vietnam and much of Laos. Its plumage and structure are similar to those of the Yellow-vented Green Pigeon, *T. seimundi*, an endemic to Thailand, Peninsular Malaysia and Indochina found in similar habitats. Both these species have wedge-shaped tails, but the Pin-tailed has longer elongated central tail feathers, is paler green overall and lacks the Yellow-vented's white belly, mostly yellow undertail region (mottled chestnut in the Pin-tailed) and maroon shoulder patch (the latter is only found in the male Yellow-vented). The two species can be found together, and often gather in flocks at fruiting trees with barbets and other frugivores.

Dave Farrow

MOUNTAIN IMPERIAL PIGEON *Ducula badia* 43–51cm

Although this is a widespread and locally common resident of evergreen forests from foothills up to the highest elevations, as is the case with all flocking pigeons, its numbers are suppressed in many areas due to hunting pressure. Its brownish-maroon upperparts contrast with the pale grey head, neck and underparts. It is slightly larger than the similarly plumaged Green Imper-ial Pigeon, *D. aenea*, from which it is distinguished by its pale vent (chestnut on the Green) and pale terminal tail-band that is conspicuous in flight. The Green is found in a variety of forest types in the lowlands and foothills, being especially numerous in the deciduous dipterocarp woodlands of Cambodia's northern plains, but it has been eradicated from most of Laos. Both species have very deep, disyllabic and far-carrying call notes (for example, *wurr-hooo'*), with the Mountain's usually preceded by a short, soft note.

Dave Farrow

BENGAL FLORICAN *Houbaropsis bengalensis* 66–68cm

Allan Michaud

This charismatic bird of lowland grasslands is Critically Endangered because its habitat is disappearing very rapidly due to both agricultural intensification and scrub encroachment. The largest remaining population in the world occurs in Cambodia's Tonle Sap floodplain, where the bird breeds in grasslands and fallowed traditional deepwater rice fields during the dry season (February–May). This is the best time to see the bird, when the males in their black and white breeding attire perform elaborate undulating aerial and neck-plume ruffling ground displays at exploded leks during the early morning and late afternoon. The females are intricately marked with brown and buff, making them very cryptic and hard to see. During the wet season the birds migrate outside the floodplain and are extremely hard to find in grassy meadows within deciduous dipterocarp woodland.

SARUS CRANE *Grus antigone* 152–156cm

Peter Davidson

The tallest bird in the region, this cultural icon is depicted on the bas-reliefs of Angkor Wat. An isolated population of 800–1,000 birds inhabits Cambodia and adjacent southern Vietnam and Laos. This Vulnerable species has declined rapidly due to widespread conversion of its wetland habitats, egg collection, capture for display and the effects of pollutants. It builds huge floating nests in the wet season in flooded meadows in Cambodia's northern and eastern plains, and migrates to traditional areas of the Mekong and Tonle Sap floodplain, like Ang Trapeang Thmor near Siem Reap, where it forms large flocks during the dry season. Pairs perform dancing courtship displays, and flocks often fly in V-formation. Their loud rolling, trumpeting calls can be heard at long range.

SLATY-BREASTED RAIL *Gallirallus striatus* 26–31cm

Dave Farrow

A locally common resident, this is the most widespread of the rails in the region, found in a variety of wetland habitats, including streams, wet rice paddies and mangroves, mainly in the lowlands. It has a reddish bill (with dark tip) and distinctive head pattern, comprising chestnut crown and nape contrasting with grey face and breast. The upperparts are dark with narrow whitish barring and spotting, and the flanks are boldly barred blackish and white. The bird wades stealthily, often cocking its short tail. In flight, its long legs trail behind its body.

PURPLE SWAMPHEN *Porphyrio porphyrio* 29–42cm

Swamphens are very distinctive birds, heavy and rather clumsy, with bright purple-blue and dark turquoise body plumage, white undertail coverts (which flash as the bird walks and cocks its tail upwards), and large red bills, frontal shields and legs. The Purple Swamphen is a locally common, fairly widespread resident of lowland freshwater lakes, marshes and floating mats of vegetation. It can be rather shy, and in densely vegetated wetlands its various grunting, chuckling and cackling sounds are all you will experience of its presence. It is hunted in some areas for its meat, and in some areas has been much reduced, but it remains extremely common around the Tonle Sap Lake.

James Eaton/Birdtour Asia

COMMON GREENSHANK *Tringa nebularia* 30–34cm

Rob Tizard

Common and widespread as a winter visitor and passage migrant, this graceful medium-sized shore bird visits both coastal and inland wetland habitats. Its breeding dress (acquired from March) differs from the non-breeding plumage (depicted) in having black mottling and notching above and heavier streaks on the neck and breast. In flight it shows a white 'V' up the back, and typically calls with a strident, trisyllabic '*tu-tu-tu*'. Several similar-looking *Tringa* species are also winter visitors, passage migrants to the region. Nordmann's Greenshank, *T. guttifer* (Endangered), a rare visitor to estuaries such as Xuan Thuy Nature Reserve, has shorter and yellower legs, a bicoloured bill and a paler head; the Common and Spotted Redshanks, *T. totanus* and *T. erythropus*, have red legs and bill bases; the Wood Sandpiper, *T. glareola*, is noticeably smaller with darker, whitish-speckled upperparts.

COMMON SANDPIPER *Actitis hypoleucos* 19–21cm

Rob Tizard

This widespread winter visitor and passage mig- rant is found in a manner of wetland habitats, from tidal creeks and rocky shorelines to rice paddies, riverbanks and lakeshores, and even urban reservoirs and ornamental water features. It has a comparatively simple plumage for a shore bird, consisting of plain brownish upperparts, white underparts, and brownish breast sides separated from the shoulder by prominent white patches, which combined with its distinctive habit of bobbing its rear body persistently make identification straightforward. It usually flies low over water with flicking beats of bowed wings interspersed with glides, often giving its characteristic shrill, piping '*tsee-wee-wee*' call. The similar-looking Green Sandpiper, *Tringa ochropus*, is larger with blackish-brown upperparts.

SPOON-BILLED SANDPIPER *Calidris pygmaeus* 14–16cm

A regular but rare winter visitor to intertidal mudflats, saltpans and shrimp ponds of the Red River Delta in northern Vietnam (although it could occur in similar habitats anywhere in the region), this is one of the world's rarest shore birds. It is Endangered through loss of its wetland breeding habitat in the Russian Far East. Both its breeding and non-breeding plumages closely resemble those of the Red-necked Stint, *C. ruficollis*. In non-breeding dress it is distinguished by its extraordinary spatulate bill (hard to see in profile), larger head, and whiter forehead and breast, creating a paler overall

Dave Farrow

appearance. It sometimes feeds by sweeping its bill from side to side. The best place to find it is among the flocks of Red-necked Stints and other *Calidris* sandpipers at Xuan Thuy Nature Reserve.

RED-NECKED STINT *Calidris ruficollis* 14–16cm

Martin Hale

he most numerous of the *Calidris* sandpipers in the region, this a locally common winter visitor and passage migrant chiefly to oasts, favouring estuarine mudflats, saltpans and brackish ponds, nd sometimes occurring on large rivers and lake shores. Many of he *Calidris* sandpipers look alike, differing only in subtle details of tructure and plumage. This is one of the smallest, with a short, traight black bill and black legs. In winter plumage it has pale grey pperparts, and whitish underparts and supercilium. In March and pril it develops the rufous head, neck and breast colour of its reeding plumage, and black-centred brown upperpart feathers. eeding behaviour and habitat are useful pointers for identifying ecies in the genus *Calidris* – this species often forms large flocks and eds with a rapid pecking action.

GREATER PAINTED-SNIPE *Rostratula benghalensis* 23–26cm

Rob Tizard

This is a widespread resident of marshy areas and rice paddies. The roles of the sexes in this unusual family (of just two species worldwide) are reversed. The female (depicted) is larger and brighter than the male, and she often mates with more than one male. The male incubates the eggs and cares for the young. This species is crepuscular, secretive and hard to see, often standing motionless for long periods. Its structure is snipe-like, but it has a droop-tipped bill, broad, rounded wings and long, trailing legs in flight. The male has buffish spectacles, and buffish mottling on folded wings. The female performs 'roding' display flights low over the ground, and her territorial call, usually heard at dusk or at night, is a long series of 'kooh' notes, like the sound made by blowing across a bottle top.

PHEASANT-TAILED JACANA *Hyrophasianus chirurgus* 29–31cm

Martin Hale

This is one of two jacana species in the region, both noted for their extremely long toes, which enable them to walk over floating vegetation on lakes and ponds. Jacanas are polyandrous: females often mate with more than one male, leaving the male to tend the eggs and rear the young. The Pheasant-tailed Jacana is a locally common resident, winter visitor and passage migrant. Shown here in non-breeding plumage, in breeding dress it has long, down-curved tail feathers, a blackish body and contrasting white wings. Its congener the Bronze-winged Jacana, *Metopidius indicus*, is a common resident (absent from central-northern Vietnam), with a plumper body, stouter yellow bill and brown wings in flight.

GREAT THICK-KNEE *Esacus recurvirostris* 49–54cm

Rob Timmins

Crepuscular and nocturnal, this rather bizarre-looking shore bird spends much of the day hidden. It has declined and is now a scarce resident of sand and gravel bars along large lowland rivers, confined to the Mekong River and its tributaries in Cambodia and Laos, as well as sand dunes, coastal flats and saltpans in central-southern Vietnam. It is wary and usually takes flight on powerful, stiff wingbeats, well ahead of an approaching observer. Its territorial call is a series of upwards-inflected, wailing whistles. Its closest relative in the region, the Eurasian Thick-knee, *Burhinus oedicnemus*, has similar habits and voice, but has a much smaller bill, and streaked breast and upperparts. The Eurasian Thick-knee has a similar regional range and frequents similar habitats, but is also found in dry cultivated, barren and scrubby areas.

PACIFIC GOLDEN PLOVER *Pluvialis fulva* 23–26cm

Dave Farrow

This medium-sized, slender shore bird is a locally common and widespread winter visitor and passage migrant found in cultivated lowlands, dry areas and coastal habitats (especially mudflats). Like other plovers, it forages by making short runs, pausing and stooping to take prey. In non-breeding plumage the bird's golden-buff-spangled upperparts, head and neck are distinctive. In breeding plumage attained in March–April), it develops black underparts bordered with white line from the forehead to the flanks. It forms flocks, in flight relayed by shrill 'chew-eet' or 'klu-ee' whistled calls. The similar Grey Plover, *P. squatarola*, is chiefly coastal and stockier, with a stouter bill; shows black armpits in flight, and has a more mournful, trisyllabic 'lee-oo-ee' call note.

47

KENTISH PLOVER *Charadrius alexandrinus* 15–17cm

Rob Tizard

This is a local resident of the coast of central Vietnam, and a widespread winter visitor to coastal habitats and large river and lake margins in the region. In non-breeding plumage (depicted) it resembles the larger, non-breeding Greater and Lesser Sandplovers, *C. leshenaultii* and *C. mongolus*, both of which are winter visitors and passage migrants to predominantly coastal habitats that lack a white nuchal collar. In breeding plumage, the male Kentish Plover has a white forehead and supercilium, black forecrown, line through eye and breast side, and rufous-buff hindcrown and nape. The male performs a stiff-winged rocking display flight. The similar but more delicate Malaysian Plover *C. peronii*, a rare resident of sandy, shell and coralline beaches in the south, has scaly upperparts. The closely related Little Ringed Plover, *C. dubius* (a locally abundant resident and common winter visitor), has a black breast band and face-mask, and a yellow eye-ring, but does not have a white wing-bar.

ORIENTAL PLOVER *Charadrius veredus* 22–25cm

Breeding male

Non-breeding

Peter Davidson

Stopping over on an exceptionally long migration from its Australian wintering grounds to its breeding grounds located in Siberia, Mongolia and northern China, this rather tall, slender plover with a long neck, wings and legs occurs only as a rare passage migrant to the region in February to April. Some birds arrive in breeding plumage, sporting chestnut and black breast bands, and a whitish head and neck; the pattern is more diffuse in females. The non-breeding plumage plainer sandy-brown. Flocks visit the short grasslands and ricefields of the outer Tonle Sap floodplain in Kompong Thom province mos. springs. This species also occurs in dry coastal habitats, for example Xuan Thuy Nature Reserve.

RIVER LAPWING *Vanellus duvaucellii* 26–28cm

A resident of large, sandy lowland rivers and surrounding areas, this striking plover is locally common along the Mekong River and larger tributaries like the Sesan and Srepok Rivers, but was once more abundant and has recently disappeared from more heavily settled tributaries. In flight it shows a tricoloured upperwing pattern that consists of a sandy-brown inner wing and black outer wing separated by a white bar; the underwing is white with black tips. The bird is very vocal, especially near the nest, when it calls with a loud and shrill '*kick, kick, kick*', repeated frequently. Its elaborate courtship display consists of stooping, spinning and stretching, which all takes place on the ground.

Rob Tizard

GREY-HEADED LAPWING *Vanellus cinereus* 34–37cm

This large plover is a widespread and locally common winter visitor from Japan and north-east China (where it breeds), favouring marshes, paddies, rice stubble and other cultivated areas, chiefly in the plains. In flight, it has a similar tricoloured wing pattern to the Red-wattled and River Lapwings, *V. indicus* and *V. duvaucellii*, but shows more extensive white in the upperwing. Juveniles differ from the adult (depicted) in having a brownish head and buff scalloping on the upperparts, and usually lack the blackish breast band. They are best told from non-breeding/juvenile Oriental Plovers, *Charadrius veredus*, by the black-tipped yellow bill and lack of white forehead/supercilium. This species tends to be much less vocal than resident (territorial) lapwings.

Rob Tizard

RED-WATTLED LAPWING *Vanellus indicus* 32–35cm

Allan Michaud

The most common lapwing in the region, and the most widespread (though nearly eradicated from some large areas, such as northern Laos), this species is found in a wide variety of wetland, agricultural and human-modified habitats, and in clearings in deciduous woodlands. One of its most memorable features and community functions is as a ceaselessly vigilant sentry by day and night – it is invariably the first to detect predators or intruders and to raise the alarm with its persistent and penetrating '*did he do-oo it*' call. In flight its wing pattern is very similar to that of the River Lapwing, *V. duvaucellii*, but the Red-wattled Lapwing has a white-tipped tail. It is usually found in pairs or trios, running about in short spurts and dipping forwards obliquely to glean prey, the classic feeding behaviour typical of most plovers.

SMALL PRATINCOLE *Glareola lactea* 16–19cm

Rob Tizard

Pratincoles are shore birds that are specially adapted to forage for insects on the wing. This is the smaller of the two pratincoles in the region, and it is a gregarious species that is very much tied to the Mekong Basin's large lowland rivers. It occurs here throughout the dry season along stretches with large sandy bars and banks, and also sometimes on dry lake margins, but where it goes in the wet season (June–October) remains completely unknown. It is faring much better than most other shore birds tied to sandbar habitats, although it is not clear why. It has a plain sandy-grey body plumage at rest that bursts into a striking black and white pattern when it takes flight. It is most active towards dusk, when flocks hawk insects on the wing with a characteristic buoyant and graceful flight action recalling the terns, as does their usual flight call. Loose nesting colonies are formed on sandbars.

ORIENTAL PRATINCOLE *Glareola maldivarum* 23–24cm

Peter Davidson

Chiefly occurring in open country between February and July, this pratincole is a breeding visitor and passage migrant to the region, especially to fallowed ricefields, dry paddies and other cultivated areas, but also to wetland fringes and coastal areas. Its warm greyish-brown and buff plumage disguises it well in the agricultural fields where it nests. The bird's short, dark bill has a prominent red base and a wide 'gape' to enable it to catch insects in flight. On the wing, it has a forked black tail and white rump and contrasting brownish upperparts; the underwing is dark with a rufous-chestnut wing lining. Juveniles lack throat markings and have scaly plumage (pale-fringed feathers).

SAUNDERS'S GULL *Larus saundersi* 33cm

Craig Robson

Smaller and more compact than its relative, the more widespread and common Brown-headed Gull, *L. brunnicephalus*, this is a rare winter visitor to the coast of north-east Vietnam. It is Vulnerable, threatened by reclamation of its feeding habitat – tidal flats and salt marshes in China, Taiwan, South Korea and Japan. In winter it occurs in estuaries and on tidal flats around the Red River Delta. The Brown-headed Gull and Black-headed Gull, *L. ridibundus*, which both occur along coasts and larger freshwater wetlands (for example on the Tonle Sap Lake, where the Brown-headed is very common), have mostly red bills, more black in the wings in adult and first winter plumages (the adult Brown-headed has white windows on black primary tips), and broader black tail bands in first winter plumage.

RIVER TERN *Sterna aurantia* 38–46cm

Rob Timmins

A localized resident of the Mekong River and its larger tributaries in south Laos and northern Cambodia, this graceful tern has been much reduced in numbers and range due to predation by people and dogs, and disturbance and conversion of sandbar habitats, which it requires for nesting. In breeding plumage (depicted) it has a black cap, bright orange-yellow bill, orange-red legs and long tail streamers. In non-breeding plumage the forehead is whitish and the bill is tipped with black; the juvenile has a blackish mask, white supercilium and blackish scaling to the upperparts. The bird has a strong, buoyant flight and feeds by plunge-diving for fish, tadpoles, shrimps and other aquatic prey.

WHISKERED TERN *Chlidonias hybridus* 24–28cm

Rob Tizard

This widespread winter visitor and passage migrant occurs in a variety of freshwater and saltwater habitats, sometimes in flocks numbering thousands (for example near the north end of the Tonle Sap Lake). It is a small and compact tern, with a short, shallowly forked tail and shortish black bill. The bird's breeding plumage (attained in March–April) differs from the non-breeding plumage (depicted) in comprising smoky-grey underparts, white 'whiskers', a black cap and blood red bill and legs. The superficially similar but very rare (almost extirpated) Black-bellied Tern, *Sterna acuticauda*, has a yellow bill, blackish underparts and long tail streamers in breeding plumage.

ORIENTAL HONEY-BUZZARD *Pernis ptilorhynchus* 55–65cm

Two subspecies of this relatively large raptor occur: one is a widespread resident, the other a passage migrant and winter visitor from temperate northern climes. Both subspecies are fairly common in evergreen and deciduous forests and open woodlands. The plumage of this bird is highly variable. It is most easily recognized by its shape when in flight, consisting of a protruding head, broad wings with bulging secondaries, held level when soaring, with two to four discrete dark bands across the flight feathers, and a longish tail with (in adults) a broad, dark terminal tail band and a narrow, dark basal band, visible from above and below. The bird can form large flocks that 'kettle' on updrafts on migration, and breaks into wasp and bee nests to feed on the larvae.

Peter Davidson

BLACK KITE *Milvus migrans* 55–60cm

Peter Davidson

Both of the two subspecies of Black Kite occurring in the region (sometimes considered separate species) are undergoing serious long-term declines: *M. m. govinda* is a rare resident in Cambodia (and possibly north-east Vietnam), and *M. m. lineatus* is an uncommon winter visitor throughout. Both occur in open habitats with trees and cultivated areas, on marshes and coasts (for example limestone karst in Halong Bay), at lower elevations and on plateaux. Both have longish, shallow-forked tails that are regularly flexed in flight, and long, angular wings with rather languid wing-beats. *M. m. lineatus* has a more rufescent plumage and (generally) more extensive white patches at the bases of the primaries.

BRAHMINY KITE *Haliastur indus* 44–52cm

Rob Tizard

Formerly considerably more widespread inland, this raptor is still a locally common resident in central and southern parts of the region, particularly around wetlands in the Tonle Sap and Mekong floodplains, and along coasts. The adult's plumage of whitish head, neck and breast, bright cinnamon-rufous body and wings with blackish tips, is very distinctive. The juvenile plumage recalls the Black Kite, *Milvus migrans*, being largely dull brownish, but juveniles have more prominent buff-white patches on the undersides of the primaries, and can be distinguished by their smaller size, rounded tail (forked in the Black Kite) and broader wings. Brahminys often feed by scavenging on flotsam and jetsam on the water's surface.

WHITE-BELLIED SEA EAGLE *Haliaeetus leucogaster* 70–85cm

This very large raptor is an uncommon resident along the region's coastline and offshore islands, which is occasionally found inland along larger rivers and around lakes. Adults form permanent pairs and these inhabit territories all the year round. The adult plumage (depicted) is mainly grey and white, but the immature is largely brown, with a dark band across the tip of its off-white tail. In flight, the bird has very broad wings held above the body in a distinctive shallow 'V' when soaring, and a short, wedge-shaped tail. It performs acrobatic display flights, and its loud, goose-like honking call immediately draws attention. It feeds on fish caught with long talons, and takes offal from the water's surface.

Peter Davidson

GREY-HEADED FISH EAGLE *Icthyophaga ichthyaetus* 69–74cm

One of two fish eagles in the region, both once widespread but now much reduced, the Grey-headed Fish Eagle inhabits lakes, swamps and wide, slow-flowing rivers with plentiful large trees in the lowlands. Its congener, the Lesser Fish Eagle, *I. humilis*, inhabits forested lowland rivers in hillier country. These are similar-looking, heavily built raptors with broad wings and short tails; the Grey-headed has a white tail with a broad black tip, and the Lesser a mainly greyish tail. Although considered fish-eating specialists, a recent study around the Tonle Sap, which supports a healthy population of Grey-headed Fish Eagles, found that they prey significantly on water snakes, which are a key component of the local economy, harvested in huge quantities to supply the crocodile-farming industry.

Peter Davidson

WHITE-RUMPED VULTURE *Gyps bengalensis* 75–85cm

Poisoning by a veterinary drug ingested when feeding on cattle carcasses on the Indian subcontinent resulted in this and other Asian *Gyps* vultures (including the Slender-billed Vulture, *G. tenuirostris*) very rapidly becoming Critically Endangered at the end of the 20th century. The small population in this region (much reduced itself) numbers perhaps a few hundred and is consequently of great conservation importance. These birds live in the deciduous woodlands of Cambodia's northern plains, ranging into southern Laos and sometimes Vietnam. The adult has blackish plumage with a white neck

Allan Michaud

ruff, and white lower back, rump and underwing lining. Immatures are rather uniform mid-dark brown in colour, with paler necks, white bars across the underwing coverts and white streaks on the underparts.

SLENDER-BILLED VULTURE *Gyps tenuirostris* 80–95cm

Allan Michaud

This species is Critically Endangered for the same reasons as its smaller relative, the White-rumped Vulture, *G. bengalensis*. Its regional stronghold is the same area of wooded plains across northern and north-eastern Cambodia. This regional population has declined significantly due to declines in wild cattle and deer, as well as changes in animal husbandry, which have both reduced the availability of carcasses to feed on. The combination of sandy plumage and long, blackish neck and head in both adult and immature plumages distinguish this species from the White-rumped Vulture. Both species nest colonially in trees or on cliffs. A recently introduced programme of supplementary feeding ('vulture restaurants') in northern Cambodia presents a novel opportunity to see these extraordinary creatures up close.

RED-HEADED VULTURE *Sarcogyps calvus* 81–85cm

Previously common, this vulture, which is found in the same areas as the other species, has suffered the same fate and is now Critically Endangered, its massive recent global decline also attributed to poisoning from consuming cattle carcasses injected with the veterinary drug Diclofenac. As is the case with the Slender-billed Vulture, *G. tenuirostris*, its decline in this region has occurred over a longer period, linked to the reduced availability of animal carcasses. The adult has a white frontal ruff and flank patches; the male has yellow eyes. Immatures are browner with pinkish head and legs. Like other vultures, this one can cover hundreds of kilometres in a single day, using warm updrafts and stronger winds at higher altitudes.

Allan Michaud

CRESTED SERPENT EAGLE *Spilornis cheela* 56–74cm

This specialist reptile hunter is the region's most common and widespread medium-large raptor. It is found in evergreen and deciduous forested habitats from plains to mid-mountains. In flight, the adult is readily identified by the black-bordered, broad white band across its underwing, and black tail with broad white central band. The bird soars with wings held slightly above the body in a shallow 'V', displays with head and tail raised, and frequently calls with a plaintive *'klee-klee wip-wip'*. While perched, the adult appears large-headed when its short, blackish crest feathers are raised, contrasting with the yellow cere, facial skin and eye. The immature (depicted) has less yellow on the face and paler underparts.

SHIKRA *Accipiter badius* 30–36cm

Accipiters (sparrowhawks) are rather short, rounded-winged and long-tailed raptors that make direct flights on rapid wingbeats. This one is a widespread and common resident of both forested and open areas, from lowland to mid-elevations (absent from north-west Vietnam), and the most likely *Accipiter* to be encountered around habitation. It is relatively small, with pale blue-grey upperparts (browner in the female than the male), and dense, narrow rufous bars across white underparts. The juvenile is darker, more streaked and spotted, with a dark streak down the centre of the throat, very like the Besra, *A. virgatus*. The bird often hunts in the open or from perches, dropping down to catch large insects, amphibians and reptiles.

Male back view (above: female front view (below)

GREY-FACED BUZZARD *Butastur indicus* 41–49cm

This medium-sized raptor is a common passage migrant and winter visitor from Japan, north-east China and Russia, where it breeds. In winter it is usually encountered singly in various open forest types and countryside with scattered trees, in lowlands and hills. On migration it can occur anywhere, and often forms flocks that kettle on updrafts when moving through mountain passes (for example in Tam Dao and Tram Ton Pass near Sapa). Its structure and flight action – rapid wingbeats interspersed with glides – recall a large, long-winged *Accipiter*. From below, it has pale underwings with black primary tips, and three dark bands across its longish tail, which is often fanned when soaring. The dark (mesial) streak down the centre of the white throat is a useful identification feature even at relatively long range.

Dave Farrow

BLACK EAGLE *Ictinaetus malayensis* 69–81cm

Nowhere common and not an easy bird to find, this widespread resident forest eagle occurs from the lowland plains up to the highest mountains. Its shape in flight – long wings, with broad hands, well-defined 'fingers' and pinched-in bases, and a long tail – is a useful pointer to its identification. In addition, it soars with wings held in a 'V', and the display flight involves steep dives and inclines through a 'U'-loop. Adults are wholly blackish, while juveniles have a pale head, underparts heavily streaked blackish, and buffish underwing coverts in flight. The bird hunts by quartering low over forest canopies. The Greater Spotted Eagle, *Aquila clanga*, is similarly plumaged, but shorter-tailed, broader winged and mainly found in open habitats.

Rob Tizard

GREATER SPOTTED EAGLE *Aquila clanga* 65–72cm

Another of the globally threatened (Vulnerable) species found in the region, this is a winter visitor to open landscapes, in particular low-intensity cultivated areas, marshes, lake shores and river floodplains, breeding in northern temperate areas. Formerly common, the bird is now rare almost everywhere, except for parts of Cambodia's Tonle Sap floodplain. The combination of large size, broad 'fingered' wings, short tail, very dark brown plumage, and large bill with yellow cere separate this from other large eagles. The bird gets its name from the immature plumage, which has prominent rows of white spots on the wing covert feather tips. It holds its wings relatively flat when soaring, and may be seen almost anywhere during the spring and autumn migration.

Peter Davidson

MOUNTAIN HAWK EAGLE *Spizaetus nipalensis* 66–75cm

An uncommon but widespread resident of forested habitats in Laos and Vietnam, this large raptor is mostly found in hills and mountains, although some individuals disperse into the lowlands. Most readily identified when perched, by its unusual tall, white-tipped black crest, the even-width dark and pale tail bands (3–4 of each) are a useful supporting feature. The adult has a dark mesial (central throat) stripe and broad dark bars on the belly, the immature a streaked rear crown and nape, and pale, unmarked buffish head and underparts. The bird soars with wings held slightly forwards and above the body in a shallow 'V', but glides on level wings. It has a similar call to that of the Crested Serpent Eagle, *Spilornis cheela*, only shriller.

Craig Robson

CHANGEABLE HAWK EAGLE *Spizaetus cirrhatus* 61–75cm

A widespread resident of Cambodia, southern Vietnam and Laos, which occurs in evergreen and deciduous forests from the plains well into the mountains, this variable, slender forest eagle is unusual in having two colour morphs. It has a short and shaggy crest, like the Crested Serpent Eagle, *Spilornis cheela*. The adult pale morph (depicted) has bold dark streaks on whitish underparts and narrow, dark tail bars variable with a broad, dark terminal band; the immature has an unmarked white head and underparts. The dark morph is charcoal coloured all over but for the greyish tail with a dark terminal band. The bird soars and glides on level wings. Its call is an ascending series of shrill, far-carrying whistles.

Allan Michaud

WHITE-RUMPED FALCON *Polihierax insignis* 25–26cm

Endemic to Myanmar, Thailand and the region covered by this guide, this forest falcon is an uncommon resident of the open deciduous dipterocarp woodlands of lowlands and plateaux in the south of the region. The bird is Near Threatened because its favoured habitat is becoming increasingly degraded and disturbed. Readily identified by its smallish size, long tail and grey upperparts with contrasting white rump, the male (depicted) has a greyish head and shawl, the female a rufous head and shawl. The bird can be very inconspicuous since it often perches in the mid-canopy for long periods, making short, direct flights with rapid wingbeats. Ecologically, it is yet another benefactor of the excavating role woodpeckers and barbets play, nesting and roosting in tree-holes.

James Eaton/Birdtour Asia

PIED FALCONET *Microhierax melanoleucos* 19–20cm

Collared Falconet   *Pied Falconet*

This diminutive falcon is an uncommon resident of clearings in evergreen forest and forest edge in hills and mountains (at up to about 1,000m) in central and northern parts of both Laos and Vietnam, from where its range spreads west along the Himalayan foothills to north-east India. The similar Collared Falconet, *M. caerulescens*, is more widespread (but absent from central-northern Vietnam), smaller still, and has a white collar and forehead, and a rich rusty-chestnut throat, thighs and vent. Both of these species like to perch on exposed snags, often in small groups and for long periods, and have a rapid, direct flight action; they are also secondary cavity nesters, making use of tree-holes excavated by woodpeckers and barbets.

DARTER *Anhinga melanogaster* 85–97cm

One of only three *Anhinga* species worldwide, the Darter is resident in the south, moving seasonally in response to water levels. It is Near Threatened and has undergone widespread regional declines due to egg collection, disturbance and tree loss at breeding colonies (it nests in the late wet/early dry season), but has recently increased rapidly as a direct result of nest protection at the region's core breeding colony at Prek Toal on the Tonle Sap Lake. Cormorantesque in appearance and habits, this serpent-like swimmer's slender neck and slim, dagger-like yellowish bill are often all that show when it swims. After swimming, it often dries its plumage in the sun by spreading its wings and fanning its broad tail. Perhaps surprisingly, it also frequently soars.

INDIAN CORMORANT *Phalacrocorax fuscicollis* 61–68cm

Peter Davidson

A resident of both freshwater and saltwater wetlands and larger lowland rivers in Cambodia and southern Vietnam, this species occupies the most restricted range of the region's three cormorants. Most readily identified when seen alongside the Little Cormorant, *P. niger*, and/or Great Cormorant, *P. carbo*, the Indian is intermediate in size and proportions between the two, with a longish, slim bill, longish neck and oval-shaped head (the Great Cormorant has a large bill, thicker neck and more angular head shape). In breeding plumage it has a tuft of white feathers behind the bright bluish-green eye. In non-breeding plumage it is more extensively white on the throat, foreneck and breast than the Little Cormorant.

LITTLE CORMORANT *Phalacrocorax niger* 51–54cm

The smallest of the region's three cormorants, this locally common resident occurs in a wide variety of wetland habitats, including small ponds, large lakes and rivers, canals, coastal ponds and mangroves. Once widespread, this species is now scarce and local in Laos and in central-northern Vietnam. None- theless, it remains the most common and widespread cormorant in the region. It is distinguished from the Indian Cormorant, *P. fuscicollis*, by its short, stubby bill, shorter neck and heavy 'jowl', darker underparts in non-breeding

Rob Tizard

plumage, and the lack of the white tuft behind the eye in breeding plumage. Like other species of cormorant, the Little Cormorant nests (October–June) and roosts in colonies, and feeds by diving for fish.

LITTLE EGRET *Egretta garzetta* 55–65cm

A resident breeding population of this white heron in Cambodia and southern Vietnam is supplemented by a wintering population distributed throughout the region in all manner of inland and coastal wetland habitats, and cultivated areas. The combination of its longish, slender neck, narrow black bill and black legs with yellow-green feet helps distinguish it from other white egrets in the region. Like other egrets, it develops long white nape and back plumes in breeding dress, frequently feeds in loose aggregations, for instance across extensive wet paddies, and nests in colonies during the wet season (for example at Prek Toal on the Tonle Sap).

Pete Morris

PACIFIC REEF EGRET *Egretta sacra* 58cm

James Eaton/Birdtour Asia

There are two very different colour morphs of this coastal egret. It is a locally common resident (absent from the central Vietnamese coast) inhabiting rocky and sandy shores, offshore islands and less frequently mudflats. The dark morph (depicted) is unmistakable, but the white morph is very similar to the other white egrets, especially the non-breeding plumage Chinese Egret, *E. eulophotes* (which is Vulnerable), a rare winter visitor to tidal flats and mangroves. The Pacific Reef Egret is shorter legged (in flight its feet and only the very base of its legs project beyond its tail), and has a thicker bill with more extensive yellowish (less dark) on the upper and lower mandibles, and mostly greenish or yellowish legs.

63

GREY HERON *Ardea cinerea* 90–98cm

Allan Michaud

This familiar Old World heron is a widespread winter visitor, with a small (formerly widespread) resident population in Cambodia and Vietnam's Mekong Delta. It occurs in a wide variety of freshwater and coastal wetlands as well as cultivated areas, usually in small numbers, but hundreds-strong flocks can form in some areas (for example on the Red River at Hanoi and the Mekong at Vientiane). In flight, its dark primary and secondary feathers contrast with its pale grey wing coverts and back. Breeding adults have bright orange-yellow bills and legs. The similar Purple Heron, *A. pupurea*, a resident and winter visitor, occurs in more luxuriantly vegetated wetlands. It is a more cryptic breeder than the Grey, and resident populations have fared better; small numbers persist even in Laos, where the only breeding herons other than small bitterns are the Malayan Night Heron, *Gorsachius melanolophus*, and Little Heron, *Butorides striatus*, which have survived by being dispersed forest breeders.

GREAT EGRET *Ardea alba* 85–102cm

Martin Hale

The largest of the white egrets in the region, this is a fairly common winter visitor throughout, and a local resident on the Tonle Sap and Mekong River floodplains in Cambodia and far southern Vietnam. It has a longer neck than the other egrets, often with a pronounced kinked look. In non-breeding plumage it is very similar to the Intermediate Egret, *A. intermedia*, but is larger, and has a longer yellow bill and longer blackish legs. In breeding plumage it develops long back and short breast plumes, and during courtship its legs become reddish, its bill blackish and its facial skin bluish. It nests in colonies with other large water birds, one of the largest colonies being in the Prek Toal core area of the Tonle Sap Biosphere Reserve.

CHINESE POND HERON *Ardeola bacchus* 45–52cm

Allan Michaud

An abundant and widespread winter visitor from its Chinese breeding grounds, which narrowly extend into north-east Vietnam, this small heron thrives in anthropogenic landscapes, occurring in even the most marginal of wetland habitats like roadside ditches, intensive rice paddies and even small puddles. It has a compact, stocky structure with a short, thick neck, and white wings in flight. In non-breeding plumage (depicted) it is virtually inseparable from the Javan Pond Heron, *A. speciosa*, a locally common resident of Cambodia and southern Vietnam. In breeding plumage (attained from March) it develops a maroon head, neck and breast (cinnamon-orange in the Javan), and charcoal upperparts.

LITTLE HERON *Butorides striatus* 40–48cm

Rob Tizard

This cryptic heron is a widespread and common winter visitor, principally found along rivers and streams in forest and lake margins that are well vegetated. There are also local resident populations inhabiting mangroves, other coastal habitats and wooded banks of large rivers. The bird is smaller and more compact than the pond herons, and much darker, with greenish or yellowish legs and a mostly dark bill. The adult has slaty-grey plumage with green-tinged upperparts, black crown, long nape plumes and whitish lines on the face, throat and breast, while the juvenile is browner with heavy dark streaking on the face, neck and underparts. The plumage of the bird shown is intermediate. This species is usually solitary, shy and blends in well with its surroundings.

CINNAMON BITTERN *Ixobrychus cinnamomeus* 38–41cm

One of three smaller bitterns in the region, this is a very common and widespread resident of vegetated wetland habitats, including small urban ponds and even tall grass well away from water. Like other bitterns, it is rather shy and remains hidden for long periods, but when seen (usually in flight, with the early breeding season slow territorial flights being beautiful to watch), the adult's bright cinnamon-rufous upperparts are quite unmistakable. The juvenile (depicted) is browner, peppered with white spots above and heavily streaked below. The similar Yellow Bittern, *I. sinesis*, which is also common but less ecologically tolerant, has pale buff wing coverts that contrast with blackish flight feathers, and is generally paler yellowish-buff overall. Both species are most active near dawn and dusk, and make platform nests in reeds and scrub.

James Eaton/Birdtour Asia

WHITE-SHOULDERED IBIS *Pseudibis davisoni* 75–85cm

James Eaton/Birdtour Asia

Once a common bird in much of Indochina, including close to habitation, this medium-sized ibis is now Critically Endangered and restricted to a few areas along the Mekong River, its tributaries and adjacent wooded plains in northern Cambodia and extreme southern Laos. Hunt- ing is a contributory factor to this bird's sharp decline. The whitish neck collar of this ibis can be tinged bluish, and in flight it has a small white patch on the leading edges of the inner wings. Its calls include some quite unearthly primordial screams. Nesting pairs can be seen outside the village of Tmatboey in Cambodia's northern plains in December–March. The Mekong between Kratie and Stung Treng is another good place to look for this species.

GIANT IBIS *Pseudibis gigantea* 102–106cm

Allan Michaud

This prehistoric-looking water bird is one of the most enigmatic and sought-after animals in Indochina. Just a few hundred remain in the openly wooded plains of northern and eastern Cambodia. The mid-late dry season (January–April) is the best time to look for it – at this time it congregates around freshwater pools and meadows to probe for invertebrates. Look out for its eerie and far-carrying, bugling calls, '*Oh look, oh look..., look, look, look...*', often delivered in duet, as it leaves roost sites early in the morning. A small ecotourism venture at the village of Tmatboey in northern Cambodia has developed to accommodate birders wanting to see this and other northern plains specialities, aiming to have nature tourism sustain a component of the local economy.

BLACK-FACED SPOONBILL *Platalea minor* 76cm

A small wintering population of this bird occurs in the Red River Delta in north-east Vietnam, where it favours tidal mudflats and other coastal wetlands. It breeds on small islands off Korea and north China, and is globally Endangered because of industrial development, land reclamation and pollution throughout its range. It is usually found in flocks, feeding in shallow water by swishing its unusual bill from side to side to catch fish and crustaceans (bringing it into conflict with fish farmers on aquaculture ponds), or roosting on earth banks, when birds remain motionless, bill under wing, for long periods. The bird's all-white non-breeding plumage develops a yellow-buff crest and breast patch in breeding dress (depicted). It is only confusable with the Eurasian Spoonbill, *P. leucorodia*, a vagrant to the region, which has a pale yellowish bill-tip and lacks black on the face.

James Eaton/Birdtour Asia

SPOT-BILLED PELICAN *Pelecanus philippensis* 127–140cm

Allan Michaud

The world's largest breeding population of this Vulnerable pelican nests in flooded forest at Prek Toal on the Tonle Sap Lake (January–April), from where the birds disperse across Cambodia, southern Laos and southern Vietnam, sometimes additionally venturing further north and west. Conservation efforts (regular patrols and round-the-clock nest monitoring) at Prek Toal have reversed a long-term decline precipitated by large-scale egg collection. Dry-season fires have become a novel threat to the water-bird colonies at Prek Toal. The pelican is gregarious all the year round, and small groups convene to cooperatively feed on fish. It often soars in large flocks with other large water birds.

PAINTED STORK *Mycteria leucocephala* 93–102cm

James Eaton/Birdtour Asia

This is a common resident of wetland habitats in and around Cambodia's Tonle Sap floodplain, and the breeding colonies at Prek Toal are among the largest in the world. The bird is uncommon to rare elsewhere, although individuals and small groups disperse widely across the region in the wet season. It is readily identified by its black and white plumage, pink tertial feathers, bare orange-red head, long, pinkish-yellow, droop-tipped bill and reddish-pink legs. The juvenile is greyish-brown with blackish flight feathers and dull yellowish skin on the head. The similar Milky Stork, *M. cinerea*, is a scarce resident around the Tonle Sap and in coastal Cambodia, often flocking with the Painted; it has all-white wing coverts (these are mostly black in the Painted). Both species soar in flocks, and fly with neck outstretched, as depicted.

WOOLLY-NECKED STORK *Ciconia episcopus* 75–91cm

Large water birds are a feature of this region. This species is a rather local but not uncommon resident of various wetland habitats such as marshes, swamps, streams and pools, usually close to evergreen or deciduous forest, in southern Laos, Cambodia and southern Vietnam. Formerly much more widespread and numerous, it has undergone a dramatic decline across the region due to human persecution. It is readily identified by its glossy black plumage (purplish or dark greenish on the shoulders), white neck, black crown, grey face, white ventral region and tail, dark grey bill and reddish legs. The

Allan Michaud

juvenile is similarly patterned to the adults but duller. The Woolly-necked Stork is usually found in singles, pairs or small groups, and nests solitarily on a platform of sticks in a tall tree.

BLACK-NECKED STORK *Ephippiorhynchus asiaticus* 121–135cm

This is now the rarest of the region's storks, and mostly confined to Cambodia, where it inhabits open deciduous dipterocarp woodland with scattered pools and meadows, open floodplain and wet cultivated areas around the Tonle Sap, and sometimes mudflats in Kompong Som Bay. It is Near Threatened and increasingly rare in Asia, but stable in its population stronghold in Australia. The female has bright yellow eyes (the bird shown is a male). In flight the wings are white with a black covert bar. The juvenile has a dull grey-brown head, neck, upper body and wings, whitish underparts and lower back, and greyish or olive legs. The bird is usually solitary, occurring in pairs or trios, and rarely soars, unlike other storks.

Allan Michaud

69

LESSER ADJUTANT *Leptoptilos javanicus* 123–129cm

This is the more numerous and widespread of the two adjutants in the region. It breeds in colonies in open wooded areas near water, mainly in Cambodia, and disperses widely across the region's lowland plains, frequenting marshes, pools and flooded areas (including paddies), where it forages on fish, amphibians and reptiles. It is globally Vulnerable, the main threats to the population in southern Indochina being collection of eggs and chicks for food, and other disturbance at nesting colonies. It shares its distinctive appearance of a naked head with sparse, coarse hairs, with the Greater Adjutant, *L. dubius*. In flight, it retracts its neck (like herons) and often soars in flocks, towering out of view on warm updrafts that enable it to move vast distances very quickly.

Allan Michaud

GREATER ADJUTANT *Leptoptilos dubius* 145–150cm

This is a rare resident of Cambodia and perhaps far southern Laos – here the population of this Endangered stork represents approximately 20 per cent of the world population. The bird breeds in flooded forest at Prek Toal and in Cambodia's northern plains, dispersing to wetlands in adjacent floodplains and open woodland habitats. It shares the same conservation pressures as the Lesser Adjutant, *L. javanicus*, from which it is distinguished by its heavier bill with convex, not straight culmen, brick-pink neck and white neck ruff, long, drooping foreneck pouch (often retracted and not visible), and more bluish-grey upperparts with a pale panel in the wing, showing as a pale bar across the dark wing in flight. It often flocks with the Lesser Adjutant and other large water birds, especially when soaring.

Allan Michaud

LESSER FRIGATEBIRD *Fregata ariel* 71–81cm

Female   Juvenile

Frigatebirds are large oceanic seabirds with long, pointed wings, deeply forked tails and mostly blackish plumage. They are scarce non-breeding visitors to the coastal and offshore waters of Cambodia and southern Vietnam, most likely to be encountered in April–September along beaches exposed to the ocean, such as Sihanoukville. In this species the male is all-black with white armpits, the female has a white breast and armpits, and the juvenile has a white or rufous head and white breast. Two very similar species also occur in the region: the Great and Christmas Island Frigatebirds, *F. minor* and *F. andrewsi*. They all take food from the sea surface and pirate (klepto-parasitize) regurgitated food from other birds. They often soar for long periods, and roost communally on offshore islets.

EARED PITTA *Pitta phayrei* 20–24cm

Pittas are plump and short-tailed, spend most of their time on the forest floor, move about with hops or bounds, and are often secretive and difficult to observe. This species is an uncommon resident of evergreen and mixed deciduous forests, secondary growth and bamboo in lowlands and hills in Cambodia, Laos and northern Vietnam. It has a cryptic plumage consisting of rich brown upperparts and golden-buff underparts with black scales, and black-scaled, whitish elongate feathers over and behind the eyes that are sometimes held erect behind its head like ears. All pittas have distinct territorial calls; the Eared Pitta gives a drawn-out and airy wolf-whistle, '*wheeooo-wit*', often near dusk.

BLUE-RUMPED PITTA *Pitta soror* 20–22cm

This pitta is near endemic to Indochina (its range marginally extends into south-east Thailand and south China). It is a widespread and locally common resident of evergreen forests situated at low to mid-elevations (of up to 1,000m). Its body plumage is rather subdued, consisting of dull green upperparts, dull buff underparts and a bluish rump; the male also has a blue nape. The Blue-naped Pitta, *P. nipalensis*, looks similar, but is only found in northern Indochina and lacks blue on the rump. The Rusty-naped Pitta, *P. oatesi*, which also looks similar and occurs at above 800m in Laos and parts of Vietnam, has a rufous head and underparts, and also lacks a blue rump.

BAR-BELLIED PITTA *Pitta elliotii* 20–21cm

Another near endemic to Indochina (its range marginally extends into south-east Thailand), this pitta is truly a jewel of the forest floor. It is a widespread and locally common resident of lowland evergreen and mixed deciduous forests below 800m. Both sexes have bright green upperparts, iridescent blue tail, blackish eye-mask and yellow underparts with blackish bars. Males (depicted) have an iridescent green crown (warm buff in females) and blue central belly and vent. This bird can be secretive and hard to find, but listening for the male's distinctive whistled, trisyllabic territorial call, 'to-weee woo', between January and May is a good way of tracking down this stunning ground forager.

LONG-TAILED BROADBILL *Psarisomus dalhousiae* 24–27cm

Broadbills are sociable birds that travel through the mid-canopy in small parties. This one is a widespread and locally common resident of evergreen forest, mostly in hills and mountains (but absent from extreme southern Vietnam). Although strikingly coloured, it can be rather shy and difficult to observe. In flight it shows a white wing patch, and when perched it has the habit of frequently jerking its tail. Its distinctive call, a shrill, downwards-inflected *'pseeeyou'* repeated several times, is often the first indication of its presence. It builds a pear-shaped nest with a side entrance, suspended from a branch and usually overhanging water, and lays 5–6 eggs, which both sexes incubate.

James Eaton/Birdtour Asia

SILVER-BREASTED BROADBILL *Serilophus lunatus* 16–17cm

Rob Hutchison/Birdtour Asia

Like the Long-tailed Broadbill, *Psarisomus dalhousiae*, this species is a widespread resident that is absent from extreme southern Vietnam. It is fairly common in evergreen forest, secondary growth and bamboo, from low elevations into the mountains. It is small, compact and has a beautiful plumage pattern of soft grey, blue, rufous and blackish tones; in flight it has a broad white band across the primaries and white tips to the outer tail feathers. Like other broadbills, outside the breeding season it is usually found in small, slow-moving family groups in the mid-canopy. It has a curious, melancholy *'kee-you'* call, and also gives a staccato *'kitikitikiti...'* trill in flight.

73

BANDED BROADBILL *Eurylaimus javanicus* 22–23cm

James Eaton/Birdtour Asia

This broadbill is a fairly common resident of evergreen forests, riparian woodland and secondary growth mostly in lowlands and lower hills. The head and underparts are the same wine colour as the basic upperpart colour. The male has a blackish breast band, which is absent in the female. The Banded Broadbill inhabits the mid-canopy and canopy, feeding on fruits and fly-catching insects, for which its extraordinary bill is designed. It usually occurs in small, slow-moving family parties and can be unobtrusive, often sitting still for long periods. Its call is a whistled, falling tone 'wheeer', followed by a series of urgent rising notes. The Black-and-Red Broadbill, *Cymbirhynchus macrorhynchos*, is similar, but has a broad white streak on the wing and a bright yellow lower mandible.

ORANGE-BELLIED LEAFBIRD *Chloropsis hardwickii* 19–20cm

Dave Farrow

Leafbirds, one of only three bird families endemic to the Indomalayan Realm (South and South-East Asia), are represented by three species in this region. They have rather slender, slightly decurved bills for taking nectar and fruit and gleaning insects. This one is a common resident of evergreen forest types in foothills and mountains in Vietnam (except the extreme south) and central-northern Laos. The sexes differ, but both can be distinguished from other leafbirds by their yellowish-orange belly and vent. The female differs from the male (depicted) in having a green head and breast, and much less purplish-blue in the wings and tail. All three leafbirds have highly variable songs that can include much avian mimicry.

BROWN SHRIKE *Lanius cristatus* 19–20cm

Martin Hale

Shrikes are medium-sized birds with largish heads, stout, hook-tipped bills and long tails. This is one of the first bird species you are likely to encounter on a trip to the region in October–April. It is a very common winter visitor throughout, found in open habitats mostly at lower elevations, including parks and gardens in urban areas, cultivated areas, secondary growth and forest edge. The most common race, *L. c. confusus*, differs from *L. c. lucionensis* (depicted) in having a more chocolate brown crown and upperparts. Shrikes prey on large insects, small amphibians, reptiles, birds and mammals from exposed perches such as fence lines and the tops of bushes, and often store their food in a larder, impaling their prey on thorns or spines to consume later.

BURMESE SHRIKE *Lanius colluroides* 19–21cm

A locally common resident throughout the region, the Burmese Shrike breeds in cultivated, scrubby and open wooded habitats (especially pine and deciduous woodland) in the hills and mountains, wintering down in the lowlands, especially in deciduous dipterocarp woodlands, but also in cultivated areas. Like all shrikes, it is usually solitary or found in pairs, and is territorial even in the non-breeding season. The female is duller than the male (depicted), and juveniles look like females but have dark scales on the upperparts and underparts; this makes them easily confusable with the scarcer migrant Tiger Shrike, *L. tigrinus*, which has an unbarred pale grey crown and shawl.

Craig Robson

75

INDOCHINESE GREEN MAGPIE *Cissa hypoleuca* 31–35cm

Craig Robson

There are two 'green' magpies in the region. This one is near endemic, its range extending into south China and far south-east Thailand. Both species are fairly common residents of lowland and hill evergreen forests across much of the region, although they are absent from parts of the south, and the Indochinese Green Magpie is absent from areas in the north. The primary plumage difference lies in the tertial feather pattern, which is uniformly green in the Indochinese and black and white in the Common Green Magpie, *C. chinensis*; the Indochinese is also shorter tailed. When the birds' plumage is worn and abraded, it becomes strongly bluish. Both species are often rather shy, but have very loud and varied calls, including piercing shrill whistles and harsh scolding chatters.

GREY TREEPIE *Dendrocitta formosae* 36–40cm

This is a medium-large, long-tailed member of the crow family that is a fairly common resident in central and northern Laos and Vietnam, where it occurs in both hill and montane evergreen forests and secondary growth. The Rufous Treepie, *D. vagabunda*, looks similar, but occurs in open deciduous dipterocarp and mixed deciduous woodland in southern Indochina, is bigger, and has a black hood, rich rufous upper and underparts, silvery grey wings and a black-tipped tail. Both species travel in small, noisy flocks, sometimes joining birds like drongos and babblers. They communicate with comical fluty or metallic ringing notes and raucous chattering.

Rob Tizard

RATCHET-TAILED TREEPIE *Temnurus temnurus* 32–35cm

This treepie illustrates the diversity of the crow family in the region. It is near endemic to Indochina, elsewhere occurring only on Hainan Island (in south China) and in an isolated population on the Thailand–Myanmar border. It is partial to the wetter evergreen forests and secondary growth of lowlands and hills (at up to 1,200m) along the Annamite Mountain chain in central Laos and Vietnam. Its plumage is wholly blackish, and its long, broad tail has distinctive spikes on each feather pointing outwards, looking rather like a ratchet. Like other treepies, this one travels in small, vociferous flocks, which emit a variety of loud, ringing, raucous, rasping and squeaky calls.

János Oláh

LARGE-BILLED CROW *Corvus macrorhynchus* 48–59cm

This heavy-billed crow is an uncommon or locally common resident throughout the region, much reduced in some areas. It is found in a wide variety of habitats, from plains to high mountains, but in many lowland areas is usually associated with large, slow-flowing rivers and other open water bodies. Its distinctive longish bill has a strongly arched upper mandible. The Collared Crow, *C. torquatus*, a rare resident of north-east Vietnam, has a white collar; the

Rob Tizard

Carrion Crow, *C. corone*, a vagrant to the north, has a smaller bill. The Asian Koel, *Eudynamys scolopacea*, a large, blackish member of the cuckoo family, is closely tied to this crow, which it parasitizes by laying eggs in the crow's nests.

MAROON ORIOLE *Oriolus trailii* 24–28cm

Orioles are colourful birds of tree canopies best known for their mellow, fluty songs. Five species occur in the region, three of which are black and yellow. This one is a relatively common resident and winter visitor from southern China, found in evergreen forests and secondary growth from foothills up to the mid-mountains. Both sexes have pale irises. The female differs from the male (depicted) in having heavily dark-streaked white underparts, dull maroon upperparts and a dark brownish hood. The bird is usually found singly or in pairs, and often joins bird waves. Its song is a beautiful rich, fluty whistle of 3–4 short notes; in contrast, it has a rather rasping, drawn-out nasal call note.

János Oláh

ASHY WOODSWALLOW *Artamus fuscus* 16–18cm

Woodswallows are smooth, agile flyers and among the very few passerine birds that soar. Although resident in the region, this species is locally nomadic, probably in response to the availability of flying insects on which it feeds. It is a locally common resident throughout the region, favouring open habitats that have scattered trees and forest edge, from lowlands to mid-mountains. It has a starling-like jizz, but is stockier with a stouter, bluish bill. In flight, the broad-based, triangular wing shape, short tail and narrow white rump band are distinctive. This is a gregarious species that favours exposed perches such as wires and snags, and spends much time on the wing, soaring and gliding in loose flocks in search of insect prey.

Rob Tizard

LARGE CUCKOOSHRIKE *Coracina macei* 27–30cm

Cuckooshrikes are medium-sized, mainly insectivorous and arboreal birds, not closely related to either cuckoos or shrikes, but bearing superficial resemblances to both. This species is a widespread and common resident of deciduous, pine and open-canopy evergreen forest, and rural areas with scattered trees, from lowland plains to higher mountains. It is the stockiest cuckooshrike in the region, with a thick bill, dark face-mask, whitish vent and pale-edged, blackish primary feathers. It sticks to the canopy, utters a loud, ringing 'klee-eep' call, has a habit of flicking up or shrugging its wings alternately when alighting on a perch, and is generally encountered in pairs or small groups.

Dave Farrow

INDOCHINESE CUCKOOSHRIKE *Coracina polioptera* 22cm

This rather sluggish mid-canopy dweller is near endemic to the region, its range extending across northern Thailand into Myanmar. It is a locally common resident of deciduous dipterocarp, mixed deciduous and pine forests in the lowlands, foothills and plateaus, but is absent from the northern half of Vietnam. It looks very similar to the Black-winged Cuckooshrike, *C. melaschistos*, which occurs primarily in evergreen forests. The male (depicted) has blackish wings with whitish feather fringes (the Black-winged lacks white fringes) and a small white wing patch in flight (smaller still on the Black-winged). The female has scaled or barred underparts, a thin white eye-ring, and more extensive white undertail tips than the Black-winged. The bird's song is series of 5–7 high-pitched, whistled notes descending the scale.

Craig Robson

79

SCARLET MINIVET *Pericrocrotus flammeus* 17–21cm

Craig Robson

One of eight minivets in the region, this is perhaps the brightest, and a pleasingly common resident inhabiting evergreen and mixed deciduous forests from lowlands to mid-mountains throughout the region. The male's vivid flame-red and black plumage is like that of the Long-tailed and Short-billed Minivets, *P. ethologus* and *P. brevirostris*. The species are best separated by subtle variations in the red patterning on the black wing. Females have a near-identical plumage pattern to the males, but the red is replaced by yellow and the black by olive-grey. The Scarlet and Long-billed are gregarious, while the Short-billed typically occurs in pairs. The Scarlet calls with a repeated, piercing, whistled *'sweeep, sweeep'*, the Long-tailed with more sibilant notes and the Short-billed with very thin, sweet whistles.

BLACK DRONGO *Dicrurus macrocercus* 27–28cm

Martin Hale

Other than during migration periods, this is the only drongo in the region likely to be found in open, treeless habitats. It is widespread and abundant in open country, cultivated areas, wetlands and urban fringe. The resident population is augmented between September and April by a migratory population from China. This species feeds by sallying for insects from exposed perches, often near livestock that disturbs prey from the ground. In winter, large communal roosts occur in some areas (such as the Tonle Sap floodplain). The Spangled Drongo, *D. hottentottus*, is a widespread and common resident and winter visitor to all forest types, and is larger, with a square-ended tail that has strongly upturned corners, and a longer decurved bill.

ASHY DRONGO *Dicrurus leucophaeus* 26–29cm

Several subspecies of this common and widespread drongo occur in the region; some are resident and others are non-breeding visitors or passage migrants. All occur in forest and forest-edge habitats, from the plains up to the higher mountains. Structurally this bird is very much like the Black Drongo, *D. macrocercus*. The plumage of resident races is also like that of the Black Drongo, but greyer, with glossiness restricted to the upperparts. Some visiting races are pale ashy-grey (like the bird illustrated), with whitish patches on the face. Darker Ashy Drongos look like the Bronzed Drongo, *D. aeneus*, which occurs in similar forest habitats, but is smaller and has more iridescence. Like other drongos, the Ashy Drongo is a fine vocalist and mimic.

James Eaton/Birdtour Asia

GREATER RACKET-TAILED DRONGO
Dicrurus paradiseus 33–35cm

One of two forest-dwelling drongos in the region with extraordinarily long, pendant-tipped outer tail feather shafts, this species inhabits a wide variety of forest types chiefly in the lowlands and hills. Its congener, the Lesser Racket-tailed Drongo, *D. remifer*, occurs mainly in hill and montane evergreen forests, but locally descends to lower altitudes. The Greater is distinguished from the Lesser by its larger size, proportionately shorter, shallowly forked tail (square-tipped in the Lesser) with twisted tail-tip pendants (flattened in the Lesser), heavier bill and a tuft or crest of feathers on the forehead. The Greater is another superb mimic. Both species can be aggressive, and frequently join or lead bird waves.

János Oláh

BLACK-NAPED MONARCH *Hypothymis azurea* 16–17cm

This flycatcher-like bird is a common and widespread resident throughout the region, except far northeast Vietnam. It inhabits forests of all types, secondary growth, plantations and even wooded urban fringes, mostly in lowlands and hills. The male is bright blue with a black 'bump' on the back of the head, a narrow black breast band and white underparts; the duller female (depicted) has the blue chiefly restricted to the head. The bird's song is a clear and ringing, rapidly whistled 'wee-wee-wee-wee-wee-wee...', and one of its characteristic calls is a rasping, rather metallic-sounding 'dzwee-ick'. It inhabits the lower-middle storeys, and typically holds its longish tail rather stiffly and slightly spread out.

James Eaton/Birdtour Asia

BLUE ROCK THRUSH *Monticola solitarius* 21–23cm

Martin Hale

This is the common rock thrush in the region, occurring as a winter visitor throughout, with a resident population in the far north, and favouring open habitats at up to 1,800m, especially rocky areas, boulder-strewn rivers, towns, roadsides and coasts. The males of the two races look different: *M. s. pandoo* (depicted) is dark greyish-blue all over, scaly in non-breeding plumage; *M. s. philippensis* has a chestnut belly and vent, similar to the darker-plumaged Chestnut-bellied Rock Thrush, *M. rufiventris*, a resident of montane forests in the north. Females are uniformly scaly brownish-grey. Winter visitors often defend territories, chasing other birds and singing frequently. The bird's upright perching posture is very noticeable on rooftops and aerials in towns.

WHITE-THROATED ROCK THRUSH *Monitcola gularis* 18–19cm

This small rock thrush is a widespread but quite uncommon winter visitor to forest, forest edge, secondary growth and plantations in lowlands up to mid-elevations. It is smaller and slimmer than the Blue Rock Thrush, *M. solitarius*, with very differently plumaged sexes. The male has a blue crown and nape, blackish mask, narrow white throat, rufous-orange underparts and rump, blackish upperparts with brown scales, a blue shoulder patch and a white wing patch. The female has boldly blackish-scaled, brownish upperparts and whitish underparts, with a whitish throat (distinguishing it from the female Blue Rock Thrush). The bird feeds on the ground and in the lower storeys of forest.

Craig Robson

BLUE WHISTLING THRUSH *Myophonus caeruleus* 31–35cm

This large, robust thrush with glossy spangled plumage and a broad, rounded tail is a common resident throughout the region. It is generally found along forest streams and rocky river-beds, around cave and cliff-bases, and in the shady undergrowth of gullies and ravines, from the lowlands up to the highest elevations. Wintering populations also occur in Laos and northern Vietnam, one of which has a black bill. The bird's distinctive call is a rather piercing, harsh 'screee', very like that of the White-crowned Forktail, *Enicurus lesch-enaulti*, which occurs in similar habitats. Its song consists of a melody of sweet, fluty, and harsher, scratchy notes. It is shy, is nearly always encountered in singles or pairs, and nests in crevices and boulders near water.

Dave Farrow

ORANGE-HEADED THRUSH *Zoothera citrina* 21–23cm

Widespread but uncommon, this bright forest thrush is a resident, passage migrant and winter visitor, found in evergreen forest, secondary growth and thick scrub in lowlands and hills. The adult plumage (depicted) is an unmistakable combination of orange and blue-grey (the female is a duller olive-brown above). Juveniles are duller, with a browner head, two dark cheek bars, and dark scales on the breast and flanks. The bird is usually seen on the ground, hopping through the leaf litter and preying on invertebrates like snails and worms, but it also feeds on fruits in the canopy of trees. It is rather shy and is typically found in singles, and occasionally in pairs. The Rusty-naped Pitta, *P. oatesi*, is more robust, and dark olive above and dull rufous below.

SCALY THRUSH *Zoothera dauma* 27–30cm

An uncommon resident across the north and in southern Vietnam, and a winter visitor throughout the region, this shy thrush occurs in evergreen and mixed deciduous forests at all elevations, and can be found in almost any habitat on migration. Its boldly black-scaled plumage is unmistakable if the bird is seen well, but all too frequently encounters with it are brief. It forages on the ground, in leaf litter and shady gullies and ravines, and sings from trees. Resident birds give slow series of three- or four-note phrases interspersed with pauses, winter visitors a soft and drawn-out whistle. When flushed from the ground, the bird often calls with a short and thin 'tsee'.

DARK-SIDED FLYCATCHER *Muscicapa sibirica* 13–14cm

This bird is one of several medium-sized 'brown' flycatchers in the region, characterized by nondescript plumage, longish wings and a rather distinctive upright perching posture. It is a common winter visitor and passage migrant throughout the region, and may breed in the far north. It occurs in all forest types, mature and degraded, but prefers relatively open canopies and forest edge (for making sallies to catch insects), across all elevations, and frequently drops into parks and gardens in urban areas on migration. Its dark-smudged breast and flanks, and shorter bill, distinguish it from the Asian Brown Flycatcher, *M. dauurica*, and its dark brown plumage easily separates it from the strongly rufous-buff Ferruginous Flycatcher, *M. ferruginea* (a passage migrant and winter visitor).

Dave Farrow

ASIAN BROWN FLYCATCHER *Muscicapa dauurica* 13cm

Martin Hale

This is a widespread and very common winter visitor and passage migrant, which is found in open forest, forest edge and clearings, secondary growth, parks and gardens, from lowlands up to mid-elevations. The photograph illustrates well the features that in combination identify this species: plain grey-brown upperparts and off-white underparts, a pale eye-ring and loral patch (between eye and bill), and a pale yellowish or fleshy basal half to the lower mandible. The bird is usually solitary, and makes rapid sallies for insects from exposed perches. The Brown-breasted Flycatcher, *M. muttui* (a scarce resident), has a longer bill with entirely yellowish lower mandible and warmer body plumage.

MUGIMAKI FLYCATCHER *Ficedula mugimaki* 13cm

Peter Davidson

This is one of three black and orange/yellow-coloured flycatchers that winter in or migrate through the region, and breed in north-east Asia (the others being the Yellow-rumped and Narcissus Flycatchers, *F. zanthopygia* and *F. narcissina*). It occurs in the mid-canopy and canopy of evergreen forest, pine forest, and parks and gardens on migration, from lowlands up to mid-mountains. The adult male's plumage (depicted) is unmistakable, although the female and immature male are less startling; they have brownish upperparts, generally one or two pale bars across the wing, and an orange-washed breast and flanks, similar to female *Cyornis* flycatchers, which lack wing-bars, and to the Red-throated Flycatcher, *F. parva*, which has much more prominent white patches at the base of the tail.

RED-THROATED FLYCATCHER *Ficedula parva* 13cm

Martin Hale

The most widespread and abundant *Ficedula* flycatcher in the region, this is a winter visitor from the Siberian taiga, where it breeds. It is likely to be encountered lower down in wooded habitats, forest edge, scrub, parks and gardens, from lowland plains well up into the mountains. It is only breeding plumage males (depicted) that have a red throat, which is gained from late February onwards; females, non-breeding males and immatures have clean white throats. The birds frequently droop their wings and cock their black tails, flashing the contrasting white bases to the outer tail feathers, and often feed by dropping to the ground to pick up insects.

LARGE NILTAVA *Niltava grandis* 20–21cm

Peter Davidson

This is the region's largest flycatcher, and typical of the niltavas in being strongly sexually dimorphic. It is a widespread and fairly common resident (but absent from parts of central and extreme southern Vietnam), found in evergreen forest in hills and mountains. The male (depicted) has a shining iridescent blue crown, neck, shoulder and rump patches. The female is brownish, with a blue neck patch, pale buff throat gorget and rufous-tinged wings and tail. This is an unobtrusive bird of the middle storeys. It has a softly whistled song, a cadence of four ascending notes, which is also used at times to signal alarm (a tactic used by all niltavas). The Small Niltava, *N. macgrigoriae*, looks similar, but is considerably smaller and generally frequents dense undergrowth.

RUFOUS-BELLIED NILTAVA *Niltava sundara* 18cm

This stunning large flycatcher is an uncommon resident or winter visitor to central-northern Laos. It is very similar to the Fujian Niltava, *N. davidi* – an uncommon winter visitor to Laos and the northern half of Vietnam from its southern Chinese breeding grounds. Both species inhabit the undergrowth and lower storeys of evergreen forest in hills and mountains. The male Fujian Niltava differs from the male Rufous-bellied Niltava (depicted) in having the shining blue on the head restricted to the forehead, no shining blue shoulder patch, and paler orange belly and vent. The females are very similar, brownish overall with a white throat gorget, blue neck patch, and rufous-tinged wings and tail. Both species take insects and fruits.

Rob Tizard

PALE BLUE FLYCATCHER *Cyornis unicolor* 16–17cm

This is an uncommon resident of fairly open evergreen and mixed deciduous forest types, from lowlands up to 1,600m, in Laos, Cambodia and parts of central Vietnam. The male (depicted) appears superficially similar to the much brighter Verditer Flycatcher, *Eumyias thalassina*, a common resident of open forest and clearings at all elevations, which has a distinct turquoise tinge to its body in both male and female plumages. The female Pale Blue Flycatcher has uniform brownish-grey plumage with rufous uppertail coverts and tail. The bird is an inconspicuous inhabitant of the mid-canopy and canopy, and has a melodious, thrush-like song, beginning with shorter '*chi*' notes and ending with buzzy '*chizz*' or '*wheez*' notes.

Craig Robson

ORIENTAL MAGPIE ROBIN *Copsychus saularis* 19–21cm

Rob Tizard

One of the region's most familiar birds, the magpie robin is a widespread and common resident that thrives in human-modified habitats, including urban areas, gardens, cultivated areas, open woodland and secondary growth, although it is not clear what its natural habitat is, or was. The male's bold plumage pattern (depicted) is mirrored in the female, only in more subdued tones. Except in heavily hunted areas, males choose conspicuous perches (such as TV aerials and radio masts), to deliver their rich, warbling song, which is interspersed with harsher notes. The bird calls with a long, plaintive, rising whistle, as well as a rasping, drawn-out '*churr*'. Typically birds are very confiding, and have the habit of cocking their tails frequently.

BLUE-THROATED FLYCATCHER *Cyornis rubeculoides* 14–15cm

James Eaton/Birdtour Asia

There are three *Cyornis* ('Blue') flycatchers in the region with orangey underparts in both sexes. This one is a fairly common resident, winter visitor and passage migrant in southern Laos, eastern Cambodia and central-southern Vietnam. The male (depicted) has a blue chin and variably rich to pale orange wedge up its blue-sided throat. The female has brown (not blue) upperparts, an orangey wash across the breast (paler on the throat and chin), a whitish belly and a rufous tail. The Hill and Tickell's Blue Flycatchers, *C. banyumas* and *C. tickelliae*, look very similar, but males of both species lack blue throat sides. Tickell's favours disturbed lowland forest in central-southern parts of the region, while the Hill is a species of evergreen forest at mid-high elevations in central and northern parts of the region.

GREY-HEADED CANARY FLYCATCHER
Culicicapa ceylonensis 12–13cm

This unusual, bright and engaging flycatcher is a common resident across most of the region. It inhabits a wide variety of evergreen and deciduous forests types, from lowland plains to the highest summits. Most birds breed in hills and mountains, then large numbers move down to winter in the lowlands, including parks and wooded gardens. The species' olive, yellow and grey plumage is very distinctive, but even more memorable is its extremely active and vocal nature. It communicates with an array of loud, metallic trilling and chattering call notes, and a clear, whistled song of four or five sweet notes (for example 'whichoo-whichoo-wit'), with an upwards inflection on the final note. It is a conspicuous and frequent member of bird waves.

Rob Tizard

BLUETHROAT *Luscinia svecica* 14–15cm

This is a largely terrestrial chat that is a locally common winter visitor and passage migrant to Cambodia, Laos and parts of Vietnam, inhabiting wetland fringes, and damper grasslands, scrub and cultivated areas of the plains and foothills. The male has a stunning blue, red and black throat and breast pattern; in females the throat is whitish, encircled by a gorget of blackish streaks, with older birds sometimes showing some blue and even rufous. This species spends most of its time on the ground, frequently cocks its tail and tends to be rather skulking often foraging under bushes. Its calls include a distinctive, twangy nasal 'dzweenk' and a more typical, chat-like, hard 'tuk'.

Martin Hale

WHITE-RUMPED SHAMA *Copsychus malabaricus* 22–28cm

In some areas, trapping for the cage-bird trade has suppressed populations of this species, which has one of the most melodic voices and most varied repertoires of any Asian songbird, including being a fabulous mimic (it is one of the most common cage birds in Vietnam). Nonetheless, it is still a widespread and fairly common resident of both evergreen and deciduous forests (especially degraded areas), secondary growth and bamboo, mostly in lowlands and foothills. The female is a slightly duller version of the male (depicted) with a shorter tail; both sexes have prominent white rumps and often cock their tails, showing off their white outer feathers. This species favours undergrowth and the lower storeys and can be rather shy.

James Eaton/Birdtour Asia

90

WHITE-CAPPED WATER REDSTART
Chaimarrornis leucocephalus 19cm

This most charismatic represent-ative of the Himalayan stream fauna is the only redstart, and one of very few robins or chats in the entire region, in which the sexes are similar. It is scarce to uncommon in central and northern Laos and northern Vietnam, where it inhabits fast-flowing, rocky rivers, streams and waterfalls, from the foothills up to the highest elevations. This water redstart's stunning tricoloured plumage consists of a black body, bold white cap (crown and nape), and rich wine red underparts, rump and black-tipped tail. It is generally found on stones and boulders, mechanically pumping, fanning and cocking its tail. It nests in holes in banks or walls, under stones or among tree roots.

Rob Tizard

PLUMBEOUS WATER REDSTART *Rhyacornis fuliginosus* 15cm

Another member of the typical Himalayan stream community, this redstart is a locally common resident of rocky, fast-flowing rivers, streams and waterfalls in central-northern Vietnam and Laos, found from the foothills up to higher mountains. The sexes differ, the male (depicted) being slaty-blue with bright chestnut tail coverts and tail, while the female is browner above, with scaly grey and whitish underparts, a bold black and white tail pattern and two whitish wing-bars (formed by white tips to the greater and median covert feathers). This is a very active bird that often fans its tail and makes fly-catching sallies. It occurs in the same habitats as the White-capped Water Redstart, *Chaimarrornis leucocephalus*, but is rather more widespread.

James Eaton/Birdtour Asia

SLATY-BACKED FORKTAIL *Enicurus schistaceus* 23–24cm

Martin Hale

Forktails (the emblem of the Oriental Bird Club) are striking black and white, stream-dwelling birds with long, scissor-like (deeply forked) tails. Of the three species in this region, the Slaty-backed Forktail is the most widespread, found on streams and rivers, and around waterfalls in hills and mountains. The White-crowned Forktail, *E. leschenaulti*, the largest forktail species, with a white crown and black breast, occurs in central and northern areas; the Little Forktail, *E. scouleri*, is a Himalayan species found locally in the far north. Forktails are often rather shy, and when alarmed utter shrill and harsh whistles that recall the Blue Whistling Thrush, *Myophonus caeruleus*. They feed on aquatic invertebrates like dragonfly larvae, and their presence is usually an indicator that a stream or river is in good health.

GREEN COCHOA *Cochoa viridis* 27–29cm

János Oláh

Closely related to the chats and thrushes, this is a scarce to uncommon resident of the hill and montane evergreen forests of Laos and most of Vietnam (except the far south). Its striking plumage, comprising a blue head, black-tipped tail, green body and black wings with extensive silvery-blue patches, belies how difficult it, and its congener the Purple Cochoa, *C. purpurea* (found in northern Laos and northern Vietnam), can be to see. Both species are inconspicuous inhabitants of the mid-canopy and canopy, although they do drop to the ground to feed and often sit motionless for long periods. Both are frugivorous, and are frequently found in close proximity to fruiting trees. Listening for their songs – which consist of loud, pure monotone whistles of up to two seconds duration – is often the best way of tracking them down.

COMMON STONECHAT *Saxicola torquata* 14cm

An abundant winter visitor to the whole region, with resident populations in northern and southern Laos and northern Vietnam, this small, compact chat likes open habitats, especially paddies and other cultivated areas, grasslands and scrub, from lowland plains to higher mountains. Both the female and the non-breeding male plumages are sandy brown with dark streaking above, a plain buffish rump and underparts, washed rufous on the breast and a white wing patch. When in breeding plumage (March onwards), the male (depicted) develops a distinctive black, rufous, brown, buff and white plumage. It drops to the ground from exposed perches (for example on top of bushes and along fence lines) to feed on invertebrates.

Martin Hale

JERDON'S BUSHCHAT *Saxicola jerdoni* 14cm

This longish-tailed chat is a localized resident that breeds during the dry season on scrub-covered rock bars in the Mekong channel upstream of Vientiane in north Laos to the Chinese border (where densities are locally very high), and in upland grasslands in northern Laos and north-east Vietnam. The male's plumage (dep- icted), consisting of blackish upperparts and white underparts, is quite bold and unmistakable. The female is a somewhat plain brown, with a whitish throat, buffish underparts and a rufous rump and tail, distinguished from the similar female Grey Bushchat, *S. ferrea*, by the lack of patterning on the head. Jerdon's Bushchat is generally shyer than other chats, and when not singing, more skulking in its behaviour.

Rob Tizard

93

GREY BUSHCHAT *Saxicola ferrea* 14–15cm

Rob Tizard

This is a fairly common winter visitor and a local resident occurring in cultivated areas, grasslands, scrub, pine forest, and evergreen and deciduous forest edge, from lowland plains up to higher mountains (breeding at above 1,200m). In breeding plumage the male sports a bolder version of the distinctive grey, black and white non-breeding plumage (depicted), losing the brown tones. The female shows a similar but more subdued patterning in brownish hues (darker in breeding plumage), with a buff supercilium, dark brown face-mask, and rufous rump and outer tail, and is more boldly marked than the female Jerdon's Bushchat, *S. jerdoni*. Like other chats, this bird chooses prominent perches from which to forage.

BLACK-COLLARED STARLING *Sturnus nigricollis* 27–30cm

Martin Hale

Common and widespread, this large, very vocal starling is a resident found in open country, cultivated areas, scrub and locally urban fringe from the plains up to mid-elevations. The Asian Pied Starling, *S. contra*, an uncommon resident of Cambodia and (formerly at least) north-west Laos, looks similar, but is smaller with reddish facial skin and bill base, and black throat, rear crown and nape. The Black-collared's vocal repertoire is not as varied or mimetic as those of many other starlings, but it has one especially distinctive call that is frequently repeated – a loud, shrill, slightly hoarse 'cheeooo', drawn out and falling in tone. The bird builds an untidy domed nest, which in some areas is parasitized by the Asian Koel, *Eudynamys scolpacea*.

COMMON MYNA *Acridotheres tristis* 25–27cm

Locally common throughout the region, this is a very successful colonist of human-modified environments, which occupies open habitats, scrub, cultivated areas, rural settlements, and towns and cities in the lowlands and hills. It is a prolific vocalist, delivering a variety of whistles, chatters, gargles, creaks and scolds, and very sophisticated mimicry of other bird songs and calls. Despite their strong ties to humans, many individuals remain wary, as they are also targets for the cage-bird trade (for their ventriloquism). Persecution is seriously restraining the species from expanding in much of Laos. It is highly gregarious and roosts in noisy communal gatherings.

Martin Hale

WHITE-VENTED MYNA *Acridotheres grandis* 25–27cm

Allan Michaud

A similarly vocal, but less mimetic relative of the Common Myna, *A. tristis*, this bird may be an important indicator of regional change in rural landscapes. Present throughout the region, its status varies greatly: it is scarce and apparently declining in Vietnam, undergoing rapid recent declines in many parts of Laos, but increasing rapidly around Vientiane. It inhabits all manner of open habitats with sparse trees, both rural and urban, often near water, frequently in close association with domestic livestock, which disturbs insects on which the myna preys. It looks very much like the Crested Myna, *A. cristatellus*, which is locally common in eastern Laos (south from Savannahkhet) and Vietnam, but the Crested has an ivory (not yellow) bill, pale orange eye, and black undertail coverts with narrow white bars.

GOLDEN-CRESTED MYNA *Ampeliceps coronatus* 22–24cm

János Oláh

This black and yellow myna looks very similar to the more numerous Hill Myna, *Gracula religiosa*, which is one of the region's most popular cage birds because of its quite extraordinary parrot-like vocal capabilities. The Golden-crested Myna is an uncommon and rather local resident of evergreen and mixed deciduous forests of lowlands and foothills, but is absent from far north-west Vietnam. Its glossy black body plumage, golden-yellow crown, cheeks and throat with contrasting dark eye, and yellow wing patch (mainly seen in flight) are unmistakable. Its calls include piercing metallic whistles, harsh screeches and bell-like notes. It is gregarious, often seen in flocks perching on exposed snags. The Hill Myna is larger, and has a heavy orangey bill and a black head with yellow wattles on the cheeks and nape.

Chestnut-**VENTED NUTHATCH** *Sitta nagaensis* 13cm

The nuthatches are immediately recognizable by their ability to climb both up and (head first) down tree trunks and branches as they search the bark for their invertebrate prey, using their powerful, dagger-like bills to remove bark and break into nuts. This species is a common resident of montane evergreen, deciduous and pine forest in southern Laos, and north-west and southern Vietnam. The similar-looking Chestnut-bellied Nuthatch, *S. castanea*, has wholly rufous-chestnut underparts contrasting with white cheeks and throat; it is a widespread and locally common resident of deciduous dipterocarp, pine and occasionally evergreen forests of the lowland plains, hills and plateaux. Both species often accompany bird waves.

János Oláh

VELVET-FRONTED NUTHATCH *Sitta frontalis* 12–13cm

This delightful nuthatch is common and widespread throughout the region, and can be found in a wide variety of forest types (evergreen, deciduous and pine forests), from the lowlands up to mid-mountain altitudes. It shares its distinctive violet-blue, beige and lavender plumage with a scarce and local near endemic, the Yellow-billed Nuthatch, *S. solangiae* (Near Threatened), which lives only in montane evergreen forest in central-southern Vietnam (including the Da Lat Plateau), south-east Laos, north-west Vietnam and Hainan Island (China). Separation of the species is straightforward – it is based on the self-explanatory bill colour. Both species regularly join bird waves.

János Oláh

SULTAN TIT *Melanochlora sultanea* 20cm

This is an unmistakable large tit that is unlike any other, and has been placed in a genus all of its own. It is an uncommon to locally common resident of lowland and hill evergreen and mixed deciduous forests across most of the region, except the south. The subspecies *M. s. gayeti* found in southern Laos and adjacent central Vietnam differs from *M. s. sultanea* (depicted) in having a black, not yellow crest. The bird's song is a clean and mellow series of three to five 'cheeooo' notes, and its calls include a loud, shrill and rather squeaky 'cheery, cheery, cheery'. It is usually in small flocks, often accompanying bird waves, and typically stays in the canopy and higher branches of larger trees.

Rob Tizard

PLAIN MARTIN *Riparia paludicola* 11–12cm

Rob Tizard

For reasons that are poorly understood, but probably linked to habitat alteration and perhaps nest persecution, this is a member of the lowland river specialist bird community that has undergone widespread declines throughout the region. It is an uncommon resident of the Mekong River and its larger tributaries in Laos and Cambodia, and the Red River in northern Vietnam. The bird's rather nondescript sandy-brown plu- mage looks similar to the plumages of the Sand and Pale Martins, *R. riparia* and *R. diluta*, which are both winter visitors to the region's lakes, rivers, marshes and adjacent habitats. The Plain Martin is distinguished from the Sand and Pale Martins by its greyish throat and breast (lacking a discrete brownish band), paler rump and shallower tail fork. It nests in colonies, excavating tunnels in sandy river banks and bars.

BARN SWALLOW *Hirundo rustica* 15–20cm

Martin Hale

The most common and widespread of the region's hirundines, this is a winter visitor and passage migrant throughout the region. It is most abundant around lowland wetlands and irrigated rice paddies. It roosts in large numbers, for example in reeds and tall grass, and on power lines. These roosts are sometimes trapped to catch birds for merit release at temples and other locations. In summer, the bird breeds in the mountains of the north, usually near water. In non-breeding plumage it loses its long tail streamers, the blackish breast band and rufous-chestnut throat become duller, and white flecks may show in the upperparts. Some birds in central Vietnam have pale rufous underparts; these belong to the race *H. r. tytleri*. This species builds mud nests on buildings, and feeds on insects while on the wing.

PACIFIC SWALLOW *Hirundo tahitica* 13–14cm

In this region, the Pacific Swallow is only found along the coastline of southern Vietnam and Cambodia, where it is locally common in various habitats, particularly near to settlements. It is best separated from the similar Barn Swallow, *H. rustica*, by its more extensive rusty-chestnut forehead, throat and breast, with no dark breast band, duskier flanks, scalloped undertail coverts (well illustrated in the photograph), and more shallow tail fork with no long streamers (this latter feature is difficult to evaluate against non-breeding and juvenile Barn Swallows, which typically lack tail streamers between September and March). The Pacific Swallow also has a shriller and more explosive call than the Barn Swallow.

Dave Farrow

WIRE-TAILED SWALLOW *Hirundo smithii* 13–25cm

Dave Farrow

This African and Asian swallow is a lowland river specialist that occurs along rocky stretches of the Mekong River channel and its larger tributaries, dispersing somewhat over adjacent floodplain wetlands and cultivated areas. It is a local, uncommon resident that appears to have declined in recent decades for reasons that are poorly understood. It is distinguished from the Barn Swallow, *H. rustica*, by its gleaming white underparts (including throat, breast and underwing coverts), chestnut crown and square-ended tail with very long, narrow (wire-like) streamers, which can be hard to discern. It builds mud nests on rock faces and artificial structures like bridges. The upper Lao Mekong is a good area for this bird.

BLACK-CRESTED BULBUL *Pycnonotus melanicterus* 19cm

This is one of the most widespread and common bulbuls of disturbed forest and forest-edge habitats in the region, occurring from the plains up to mid-montane altitudes in some areas. Its plumage, consisting of a black head with a tall, erect crest, olive upperparts and bright yellow underparts, combined with a pale eye, are unmistakable. Its perky three- or four-note, whistled song (rendered '*whit whee-oo wheet*' '*whit whee-oo*' and similar) becomes a cheerfully familiar forest sound, often heard in the midday heat when other birdsong has receded. It has a rather woodpecker-like, undulating flight. The similar Black-headed Bulbul, *P. atriceps* (an uncommon forest resident in central and southern areas), lacks a crest and has a yellowish wing panel and tail with contrasting dark sub-terminal band.

János Oláh

RED-WHISKERED BULBUL *Pycnonotus jocosus* 18–20cm

Rob Tizard

This is a common resident throughout the region, although it is very locally distributed in Cambodia and drier parts of Laos. It thrives in degraded habitats, including secondary growth, clearings, scrub, cultivated areas, parks and gardens, from lowlands to mid-elevations. Its tall, erect, blackish crest and principally brown and white plumage are distinctive among the bulbuls. The superficially similar Sooty-headed Bulbul, *P. aurigaster*, has a much shorter crest and paler upperparts with a whitish rump, and populations in the southern half of the region (very common in deciduous dipterocarp woodland) have yellow undertail coverts. The Red-whiskered Bulbul is often found in flocks, sings with varied musical phrases and has a bubbly, rolling '*proop*' call note, similar to that of some bee-eater call notes.

FLAVESCENT BULBUL *Pycnonotus flavescens* 22cm

This montane bulbul is found in degraded and secondary habitats, including forest edge, clearings, scrub and tall grass. It is a common resident in northern areas and distributed patchily in the mountains of southern Laos and southern Vietnam. Its song consists of jolly, three- to six-note phrases. The Yellow-vented Bulbul, *P. goiavier*, a very common resident of Cambodia and southern Vietnam (in towns and cities, its cheery fluty song – almost a chuckle – is part of the garden soundscape), and also found in central-southern Laos, looks similar, but has brownish not olive upperparts and a prominent white supercilium flared behind the eye, contrasting with a dark crown.

János Oláh

PUFF-THROATED BULBUL *Alophoixus pallidus* 22–25cm

James Eaton/Birdtour Asia

This very common and vociferous bulbul inhabits the middle storeys of evergreen forests across much of the region, except south-west Cambodia and southern Vietnam, where it is replaced by the very similar-looking Ochraceous Bulbul, *A. ochraceus*. The Puff-throated Bulbul's crest and nondescript appearance enable relatively straightforward identification (other brown bulbuls, except the Ochraceous Bulbul, lack a crest). Other supporting features are its grey face and off-white throat, which often appear 'puffed up' due to feather ruffling. The bird is very vocal and is often first encountered by ear, calling with persistent raucous or harsh and occasionally explosive 'churrt' notes.

ASHY BULBUL *Hemixos flavala* 21cm

Over 20 species of bulbul have been recorded in the region. This one is a forest bulbul with a widespread distribution, although it is absent from far southern Vietnam and replaced in north-east Vietnam by the only other species in this genus, the Chestnut Bulbul, *H. castanonotus*, which can be found relatively commonly at Tam Dao north of Hanoi. Both species inhabit evergreen forest and forest edge, principally in hills and mountains, occasionally descending into the lowlands, and they frequent the upper-middle storeys and forest canopy. The Ashy Bulbul has a bright yellowish-green wing panel that contrasts with greyish upperparts (the subspecies *H. f. remotus* in central-southern Vietnam and southern Laos has a paler brown head and upperparts); the Chestnut Bulbul has chestnut head sides and upperparts.

János Oláh

ZITTING CISTICOLA *Cisticola juncidis* 11–12cm

This small and streaky warbler is a very common resident throughout the region, found in paddies, marshes, grasslands and scrub in the lowlands, and usually in seasonally flooded or wet areas. Its most striking feature is its song, for which it is named, a monotonous, much rep-eated 'zit, zit, zit...' or 'dzik-ik, dzik-ik...', given in a weakly undulating, wandering song flight. The bird has broad white tips and blackish sub-terminal markings to the tail feathers. Its close relative, the Bright-capped Cisticola, *C. exilis*, has a rufous-buff supercilium and nuchal collar, a sandy golden head in breeding plumage, and a long, nasal wheezing call note; it typically inhabits taller grass and scrub. Both species are brood parasitized by cuckoos, which seems extraordinary given the size mismatch.

Martin Hale

RUFESCENT PRINIA *Prinia rufescens* 11–12cm

János Oláh

Prinias are brownish-plumaged small warblers with longish or very long, graduated tails. This species falls into the group with longish tails. It is common and widespread, and occurs in a variety of habitats including evergreen, deciduous and pine forest undergrowth, forest edge, grass and scrub at low and mid-elevations. In breeding plumage (depicted) it has a greyish head resembling the head of the Yellow-bellied Prinia, *P. flaviventris*, a common bird of drier grass and scrub in the lowlands (including towns), which has a very long tail and yellow belly and vent. In non-breeding plumage its head is brownish, when it resembles the Plain Prinia, *P. inornata* (very long-tailed), the common prinia of damp grass and scrub in lowlands and hills, often near human habitation. See also the Grey-breasted Prinia, *P. hodgsonii* (below).

GREY-BREASTED PRINIA *Prinia hodgsonii* 10–12cm

Rob Tizard

A widespread and common resident of grassy and scrubby habitats, secondary growth and forest edge in the lowlands and hills, this prinia belongs to the long-tailed group of prinias. In breeding plumage (April–August) the bird has a dark greyish head with contrasting white throat and broad greyish breast band. By contrast, in non-breeding plumage its head is browner, the breast band is reduced or absent, and it can look very similar to the Rufescent Prinia, *P. rufescens*, from which it is distinguished by its short (or lacking) white superciliary line that does not extend behind the eye, less rufous-tinged upperparts and higher pitched, nervous, tittering call notes (these are more nasal and buzzing in the Rufescent Prinia). It is usually encountered in busy, highly communicative flocks.

BROWN PRINIA *Prinia polychroa* 15–18cm

James Eaton/Birdtour Asia

One of the very long-tailed species of prinia, the Brown Prinia is characteristic of grassy understoreys in open deciduous dipterocarp and pine forests of lowlands, hills and plateaux in Cambodia, central and southern Laos and central-southern Vietnam. It has greyish-brown upperparts with dark streaks, a faint pale supercilium and unstreaked underparts. The similar Hill Prinia, *P. atrogularis*, has a distinct white supercilium and a gorget of dark breast streaks, and generally occurs at higher elevations (above 900m). The Plain Prinia, *P. inornata*, is smaller than the Brown Prinia, lacks upperpart streaking, and favours grassy and scrubby habitats close to water. All three species of prinia are generally found singly or in pairs; the Brown and Hill are more skulking than the Plain.

MANCHURIAN REED WARBLER
Acrocephalus tangorum 13–14cm

Peter Davidson

Acrocephalus warblers have rather short wings, rounded tails and longish undertail coverts, and typically carry themselves in a horizontal posture, skulking in damp grass and scrub. This 'Acro' is a locally common winter visitor to Cambodia, and a passage migrant to the Laos Mekong floodplain and the Red River Delta in Vietnam. It is globally Vulnerable due to habitat loss on its breeding grounds in north-east China and Russia. Cambodia may be its wintering stronghold, where it occurs in grassland and scrub near water, sedge beds and wetlands in deciduous dipterocarp forest. It looks very similar to the Black-browed Reed Warbler, *A. bistrigiceps*, but is longer-billed and longer-tailed, and has less contrasting black brows. It calls with a distinct 'churrr', more gentle and drawn out than the Black-browed's short 'chuk' notes. It remains on the wintering grounds into May, when it often sings.

104

COMMON TAILORBIRD *Orthotomus sutorius* 11–13cm

Martin Hale

Tailorbirds get their name from the method they use to construct nests, which involves weaving one, two or three leaves together with plant fibres or silk from spider webs to make a cradle within which the cup-shaped nest is built. Three species of these small, long-tailed warblers occur in the region. The Common Tailorbird occurs throughout and is abundant in urban and rural areas that have been settled by humans; the Dark-necked Tailorbird, *O. atrogularis*, occurs in evergreen and mixed deciduous forests at lower elevations, and the Mountain Tailorbird, *O. cuculatus*, occurs in montane evergreen forest. All three species frequently cock their tails. The Common Tailorbird's explosive song of repeated *'chee-yup'*, *'pitch-it'* and similar phrases is perhaps the most familiar birdsong of urban and rural gardens.

DUSKY WARBLER *Phylloscopus fuscatus* 12cm

Martin Hale

The most abundant of the subgroup of brown 'Phylloscs' in the region, this is a widespread winter visitor. It is found in low bushes and undergrowth in a variety of settings, from plains to mid-mountains, often near water. Its upperparts are more greyish than those of other brown Phylloscs, and it has dirty whitish or buff-toned underparts and a prominent buffish-white supercilium. The similar-looking Radde's Warbler, *P. schwarzi*, is stockier, with a thicker bill and legs, and a broader fore-supercilium, often rich buff distally; it frequently shows bright rusty-buff undertail coverts contrasting with paler underparts. The Buff-throated and Yellow-streaked Warblers, *P. subaffinis* and *P. armandii*, both winter visitors to degraded and cultivated mountain habitats in the far north, have yellow-sullied underparts.

LEMON-RUMPED WARBLER *Phylloscopus chloronotus* 10cm

James Eaton/Birdtour Asia

This is one of a large group of green *Phylloscopus* warblers – the 'little green jobs' of the Asian bird world – requiring a careful and stepwise approach to identification; if you want to try and unravel the conundrums of this difficult group, purchasing Robson is an important first step. The Lemon-rumped Warbler typifies the subgroup with a crown stripe, wing-bars and a pale rump. Together with the Chinese Leaf Warbler, *P. sichuanensis*, which looks almost identical, it is an uncommon to fairly common winter visitor to northern parts of the region. You are much more likely to encounter the Yellow-browed Warbler, *P. inornatus*, a widespread winter visitor to parks, gardens, forest edge and secondary growth throughout the region – it gives a strident, high-pitched '*tsu-eet*' call, and lacks the pale rump and crown stripe.

BLYTH'S LEAF WARBLER *Phylloscopus reguloides* 12cm

James Eaton/Birdtour Asia

This bird is a common winter visitor to evergreen forests in central-northern parts of the region, with resident populations in the mountains of south-east Laos, as well as southern and north-west Vietnam. It is one of the subgroup of 'Phylloscs' with wing-bars and a crown stripe, but lacking pale rumps. Its bright olive-green upperparts, double wing-bar, bold yellowish-white supercilium and central crown stripe, and whitish underparts look very similar to those of the White-tailed Leaf Warbler, *P. davisoni*, a common resident in the north and mountains in south-west Cambodia and southern Vietnam, which is generally yellower on the head and underparts with more white in the tail. Blyth's Leaf Warbler tends to prefer the canopy and frequently behaves like a nuthatch, hugging trunks and branches. On its territories, this species flicks each wing alternately in an agitation display.

SULPHUR-BREASTED WARBLER *Phylloscopus ricketti* 11cm

This 'Phyllosc' is the brightest of the subgroup with wing-bars and crown stripes but lacking pale rumps. It is a locally common winter visitor and passage migrant found in evergreen and mixed deciduous forests of lowlands and hills. It has bright yellow supercilia and a central crown stripe contrasting with a bold blackish eye and lateral crown stripes, narrow yellow wing-bars and bright yellow underparts with no white in the tail. A recently discovered resident population of the Sulphur-breasted Warbler in the limestone karst of central Laos and central Vietnam is probably an undescribed sibling species, with a very similar plumage but distinctly different song, recalling that of the Yellow-vented Warbler, *P. cantator* – a winter visitor and probable resident in northern Laos that looks similar again but has a white lower breast and belly.

Peter Davidson

BIANCHI'S WARBLER *Seicercus valentini* 12–13cm

Bianchi's Warbler is one of a complex of five extremely similar species, all of which were formerly called the Golden-spectacled Warbler. Four of them occur in the region (the other species are the Plain-tailed, Grey-crowned and Marten's Warblers, *S. soror*, *S. tephrocephalus* and *S. omeiensis*). Robson is the tool for identifying this group. Bianchi's occurs in hills and mountains of northern Vietnam and northern Laos; the Plain-tailed is a winter visitor to Cambodia and southern Vietnam; the Grey-crowned is resident in far north-west Vietnam and a local winter visitor to the north and south of the region, and Marten's is a winter visitor to northern Laos. Each species inhabits evergreen

James Eaton/Birdtour Asia

forest and secondary growth. The birds' plumages are characterized by yellow eye-rings, grey crowns with dark lateral crown stripes, green upperparts, bright yellowish underparts, variable white in the outer tail and (often) a single wing-bar.

GREY-CHEEKED WARBLER *Seicercus poliogenys* 9–10cm

James Eaton/Birdtour Asia

Similar to the 'Golden-spectacled Warbler' complex, this species together with the White-spectacled Warbler, *S. affinis*, presents an identification challenge not unlike that of the *Phylloscopus* warblers. An eye for plumage subtleties and a well-trained ear for vocal differences are again the keys, combined with knowledge of distribution (have a copy of Robson handy). Both species are evergreen forest dwellers and regularly join bird waves. The Grey-cheeked is a locally common resident in eastern Laos and Vietnam and has grey lores (feathering between the bill and eye), while the White-spectacled is a local resident in southern Vietnam and perhaps a winter visitor to the far north, and has yellow lores.

SPOT-BREASTED PARROTBILL *Paradoxornis guttaticollis* 18–22cm

János Oláh

This is the largest of the region's parrotbills, and its scientific name reflects the original uncertainty over their taxonomic relationships (which remains a focus of debate today). It is a common resident of scrub, tall grass, swidden agriculture and forest-edge habitats in montane areas of northern Laos and Vietnam. Its chunky, bright yellow, parrot-like bill, the large, crescent-shaped black patch on its ear coverts, and the arrow-shaped black spots on its throat and breast are unmistakable. It communicates with a rich variety of sounds, the song typically consisting of a loud series of 3–7 short, musical and metallic notes, and calls including harsher and less strident notes. It is often found in small, roving flocks, foraging busily in thicker patches of vegetation.

YELLOW-EYED BABBLER *Chrysomma sinense* 17–19cm

A locally common resident of Laos and parts of Vietnam, this long-tailed babbler is found in scrub, grass and secondary growth from the plains into the mountains. It is easily recognized by its combination of long tail, bold blood-orange orbital rings surrounding its yellow eyes, plain rufous-brown upperparts, clean white throat, breast and short supercilium, short but strong blackish bill and straw-coloured legs. It can be very skulking, and is usually encountered in small groups. The bird's song consists of varied melodious short and strident phrases regularly repeated after brief intervals, and calls with trilling 'churr's and explosive chatters.

Dave Farrow

ORIENTAL WHITE-EYE *Zosterops palpebrosus* 11cm

White-eyes are small, gregarious, active and warbler-like birds, and they are represented in this region by three species that look similar. The Oriental White-eye is the most widespread resident, found in forests, secondary growth, cultivated areas, parks, gardens and even city centres, up to mid-elevations. The Japanese White-eye, *Z. japonicus*, breeds in the north of the region, but is mostly a winter visitor to Laos and central-northern Vietnam. It lacks the yellow ventral stripe connecting the breast and vent in the Oriental White-eye, a feature that is barely discernible in some Oriental White-eyes. The Chestnut-flanked White-eye, *Z. erythropleurus*, a winter visitor (extending south to Cambodia and southern Vietnam in some years),

János Oláh

is best identified by its rusty flanks; it often mixes with flocks of Japanese White-eyes. All white-eyes have distinctive soft but ringing, sibilant calls, repeated persistently by flocks.

BLACK-CHINNED YUHINA *Yuhina nigrimenta* 11cm

James Eaton/Birdtour Asia

Yuhinas are small, gregarious and arboreal babblers of the Sino-Himalayan zone, characterized by their pointed crests; five species occur in Indochina. The Black-chinned Yuhina is a locally common resident of evergreen forests and secondary growth in the hills and mountains, patchily distributed from north-east Cambodia and southern Vietnam north to the Chinese border. It is one of the smaller and shorter-tailed yuhinas; its mostly red bill and black face contrast with its greyish head and 'shawl', giving it a distinctive look. Like other yuhinas, it can be found in flocks that move swiftly through the middle storeys, and often joins mixed-species foraging flocks.

STRIATED YUHINA *Staphida castaniceps* 12–14cm

János Oláh

This yuhina is a common resident of evergreen forests, secondary growth and (occasionally) scrub at mid-elevations in Laos and central and northern Vietnam. It is easily identified by its combination of a short, erect pale grey crest, dull chestnut-coloured ear coverts and nape with bold white streaks, long tail with white tips to the undersides of the outer feathers, clean white underparts and dull brownish upperparts with faint white feather-shaft streaks on the mantle. Like the Black-chinned Yuhina, *Y. nigrimenta*, it is usually encountered in vociferous, chattering flocks (sometimes large), frequently mixed with other species, and descends to lower elevations in the winter.

MOUNTAIN FULVETTA *Alcippe peracensis* 14–15cm

János Oláh

The Mountain Fulvetta is one of the drably plumaged and similar-looking fulvettas (the nun-babblers), four of which occur in the region, all inhabitants of evergreen forest, secondary growth and bamboo. The Mountain Fulvetta, a near endemic, occurs above 900m in central-southern parts of both Laos and Vietnam. The Grey-cheeked Fulvetta, *A. morrisonia*, has buff-washed underparts and replaces the Mountain Fulvetta in hills and mountains to the north. The Black-browed Fulvetta, *A. grotei*, is endemic to the lower slopes of the Annamite Mountains (below 1,000m), and lacks the prominent eye-ring. The Brown-cheeked Fulvetta, *A. poioicephala*, which has a brown, not grey face, occurs in lowlands and hills in northern Laos and north-west Vietnam. All the species inhabit the lower-middle storeys, form vocal, often inquisitive flocks, and have sweet cheery songs.

RUFOUS-THROATED FULVETTA
Schoeniparus rufogularis 12–14cm

James Eaton/Birdtour Asia

The fulvettas are a diverse group of small but sturdy babblers with relatively stout bills; the Rufous-throated is one of the most attractive. It is a locally common resident of northern and central parts of Laos and Vietnam, found in evergreen forests at lower elevations, generally below 900m. It is a stunning little bird with a head pattern of chestnut, black and white, but it can be very hard to see at all, being shy and usually sticking to dense undergrowth. Its song is a useful give-away – it utters loud, shrill, sweet phrases of 4–7 notes, repeated at regular intervals of ten seconds or so.

SPOT-NECKED BABBLER *Stachyris striolata* 16cm

James Eaton/Birdtour Asia

A fairly common resident of evergreen forest, secondary growth and scrub of lowlands, hills and lower mountains of northern and central regions of both Laos and Vietnam, this chunky babbler is rather skulking and generally keeps to the undergrowth. When seen well its identification is relatively straightforward: it is structurally robust, and has a stout grey bill, a distinctive head pattern dominated by a white throat and cheek bisected by a bold black moustachial stripe, a prominent white-streaked supercilium and bold white streaks on its neck sides. Its body is brown with a strong rufous tinge to the underparts. Its song consists of three well-spaced, whistled notes, rendered '*too, tee, tee*' or '*too, tee, too*', and it typically calls with scolding '*churr*'s and rattles.

GREY-THROATED BABBLER *Stachyris nigriceps* 13–14cm

This common resident of evergreen forest and secondary growth of the lowlands, foothills and mountains in Laos and Vietnam has yet to be recorded from Cambodia, but probably occurs in the north-east. It superficially resembles the Spot-necked Babbler, *S. striolata*, but is smaller and less robust, with pale eyes, and a grey (not white) throat and cheeks bisected by a broad white moustache (but note, central-southern Vietnamese birds lack this feature). Its white supercilium is narrow and bordered above by a black line, and it has a black- and silver-streaked forecrown. It is less skulking than the Spot-necked Babbler, and frequently joins other species to form mixed feeding flocks in the lower-middle storeys.

James Eaton/Birdtour Asia

SOOTY BABBLER *Stachyris herberti* 18cm

Endemic to the seam of limestone that spans the Annamite Mountains in central Laos and Vietnam, this is a robust and enigmatic, medium-sized babbler with an unmistakable wholly sooty, brown-black plumage, contrasting pale bill, and broad pale ring surrounding a dark eye. It favours the lower storeys and undergrowth of evergreen forest on limestone, often close to rocks and boulders, and forages in small, usually con-specific flocks. Two of the most accessible locations to find it are around the bases of wooded limestone outcrops alongside the highway between Thakek and Lak Xao in Laos, and at Phong Nha Nature Reserve in Vietnam, where this bird was photographed.

James Eaton/Birdtour Asia

STREAK-BREASTED SCIMITAR BABBLER
Pomatorhinus ruficollis 17–19cm

Rob Tizard

One of six scimitar babblers in the region characterized by their sickle-shaped bills, this species closely resembles, and interbreeds with, the White-browed Scimitar Babbler, *P. schisticeps*. The Streak-breasted occurs in north-east Laos and northern Vietnam; the White-browed replaces it across the remainder of the region to the west and south. Both species inhabit evergreen forest, secondary growth, bamboo and scrub, but the Streak-breasted is typically more montane in distribution, occurring only locally in the lowlands. The White-browed is slightly larger and longer-billed, and has unstreaked, clean white underparts (the race in the south has chestnut flanks). Both species sing with a series of 3–7 rapid, clear and mellow whistled notes, such as 'huu-huu-huu....', and call with scolding rattles.

RED-BILLED SCIMITAR BABBLER
Pomatorhinus ochraceiceps 21–23cm

Red- and Coral-billed Scimitar Babblers look similar, and their distributions overlap from southern Laos and central Vietnam north-wards, but the Red-billed extends further south into southern Vietnam and north-eastern Cambodia. In terms of appearance, the Coral-billed, *P. ferruginosus*, has a shorter, stouter bill, a broader black mask, a prominent black upper border to its supercilium, and buff or even orange-buff flanks, extending across breast and belly in some races (all-white in the Red-billed). Both species are usually encountered in small foraging flocks in the lower and middle storeys, and readily join bird waves.

János Oláh

The Red-billed's song is a three-note series of rapid, mellow whistles (similar to the songs of the White-browed and Streak-breasted), but the Coral-billed's typical song is a mellow, disyllabic whistle. Both species chatter and scold harshly.

SPOTTED WREN BABBLER *Spelaeornis formosus* 10cm

Dave Farrow

This little wren babbler is a scarce or uncommon resident of undergrowth in evergreen forest, scrub and well-vegetated gullies in hills and mountains located in central-northern Laos and parts of northern Vietnam. The bird has a longer tail than the Pygmy Wren Babbler, *Pnoepyga pusilla*, and although it is brownish overall, it is rather differently patterned. Its upperparts are peppered with tiny whitish speckles, and it has black-barred rufous wings and tail. Encountered solitarily or in pairs, it keeps to the undergrowth and ground, fallen logs and boulders, and can be very inquisitive and approachable. Its almost inaudible, high-pitched song consists of a repeated series of sibilant, tinkling phrases, for example '*tit-see-eee, tit-see-eee...*'

114

PYGMY WREN BABBLER *Pnoepyga pusilla* 8–9cm

The tiny size and incredibly short tail of this wren babbler (which appears almost tail-less) give it a very distinct jizz. It is a fairly common resident of hill and montane evergreen forest in Laos, and parts of northern and southern Vietnam. Its plumage is dark brown, with paler scaling, most pronounced on the breast and flanks, and two rows of tiny buffish spots on the tips of the wing coverts. Its song consists of three thin but loud notes, each falling in tone, delivered over 3–5 seconds and often rendered 'three...blind ...mice'. Southern Vietnamese birds sing with just two notes. It stays on or near the ground, and looks very like the Scaly-breasted Wren Babbler, *P. albiventer*, which occurs only in far north-west Vietnam and has a prominently scaled head.

James Eaton/Birdtour Asia

GOLDEN BABBLER *Stachyridopsis chrysaea* 10–12cm

Peter Davidson

This is a small and warbler-like tree babbler that is a common resident of hill and montane evergreen forests throughout Laos and central-northern Vietnam, but absent from Cambodia and southern Vietnam. Its bright golden-yellow plumage with blackish streaking and narrow black face-mask give it a very striking appearance. It is an extremely active member of the lower-middle storey bird community, usually in small flocks, and frequently joins bird waves. Its song, consisting of a series of monotone, mellow piping notes, 'tu, tu-tu-tu-tu-tu-tu-tu', is very similar to that of the Rufous-fronted and Rufous-capped Babblers, *Stachyris rufifrons* and *S. ruficeps*. These birds occupy similar ranges but typically more degraded habitats, and have duller plumages, with rufous forecrowns and yellowish-buff underparts.

GREY-FACED TIT BABBLER *Macronous kelleyi* 14cm

James Eaton/Birdtour Asia

Endemic to central and southern Laos and Vietnam, and north-east Cambodia, this close relative of the Striped Tit Babbler, *M. gularis*, which is common and widespread throughout the region, inhabits evergreen forest of the lowlands and hills at up to about 1,100m. Its plumage differs only subtly from that of the Striped Tit Babbler, being greyer (less yellow) on the face, with very sparse streaking restricted to the breast sides and more rufescent (less olive) tinged upperparts. Its song is a series of up to 20 softly whistled '*tuu*' notes, of an even tone; the Striped Tit Babbler gives a (usually shorter) series of more ringing, hollow '*chonk*' notes, which are often preceded with a short '*ti*'. It is also more generalist in its habitat choice, using a variety of degraded, scrub and grassy habitats.

PUFF-THROATED BABBLER *Pellorneum ruficeps* 16-18cm

This rather small but strong-legged babbler is quite terrestrial in its habits. It is a widespread and very common resident, found in a variety of habitats, including evergreen and deciduous forest, secondary growth, scrub and bamboo, from lowland plains up to mid-mountains. Its bright rufous cap, prominent whitish supercilium, whitish underparts with heavy dark streaks on breast and flanks, and pale fleshy legs, coupled with its habit of creeping about, rather thrush-like, on the ground where it feeds, make it readily identifiable. It is very often solitary, and has a repetitive song consisting of a 2–3-note whistled phrase, '*wee-chew*' or '*wi-tee-chew*', the last note falling in tone.

James Eaton/Birdtour Asia

SCALY-CROWNED BABBLER *Malacopteron cinereum* 14–17cm

János Oláh

A representative of a genus of babblers typical of the Sundaic region (Malaysia and Indonesia), this medium-sized tree babbler is a common resident of lowland evergreen forests in central-southern parts of this region. Its rufous crown with black scales, which can be hard to discern, bordered at the rear by a broad black nape band, are distinctive features shared with no other babblers in this region. It is usually encountered in small parties, often mixed with other species, favouring the lower and middle storeys. Its song is varied, but comprises a series of 4–6 rapid, usually either ascending or descending, short, whistled notes; several birds often sing in unison, treating the listener to quite the chorus.

ABBOTT'S BABBLER *Malacocincla abbotti* 15–16cm

James Eaton/Birdtour Asia

Rather stocky, strong-billed and shortish-tailed, this babbler is a common resident of evergreen forest, secondary growth and thick scrub of lowlands and hills in Cambodia, Laos, and parts of central and far southern Vietnam. Its plumage is a nondescript brown, with a whitish throat, rufous-tinged flanks and vent, and a greyer face and faint supercilium. Its song consists of variable cheery phrases of 3–5 well-spaced, whistled notes, one of which is often rendered *'three cheers for me'*. It spends most time in the undergrowth and is especially fond of rattans. The Buff-breasted Babbler, *Pelloreum tickelli*, is similar in appearance and occurs in the same habitats throughout the region, but it has a buffish breast with faint streaks, lacks rufous tones, and is both longer-tailed and slimmer-billed.

LIMESTONE WREN BABBLER *Gypsophila crispifrons* 18–20cm

János Oláh

Endemic to Indochina, Thailand and Myanmar, this is a locally common resident of forest on limestone at lower elevations (below 900m) in northern Laos and central-northern Vietnam. The bird's structure and behaviour recall the laughingthrushes, and it is generally encountered travelling in small parties. Its plumage is dark greyish-brown, with a subtly scaled shawl (crown and mantle), pale grey supercilium, pale throat with bold blackish streaks and greyish underparts. Its song is a rapid series of loud, slurred whistles, voluminous enough to cause surprise when bursting abruptly out of moist or shady, rocky undergrowth, where this species spends most of its time.

STREAKED WREN BABBLER *Napothera brevicaudata* 14cm

James Eaton/Birdtour Asia

Recalling the Limestone Wren Babbler, *Gypsophila crispifrons*, in its structure and appearance, but not tied to forest on limestone, and consequently very much more widespread, this species occurs throughout the region, except in far southern Vietnam. It inhabits evergreen forest, both on and away from limestone, but is most typically found where there are rocks and boulders protruding from the soil, and along dry or intermittent stream beds. It is noticeably smaller than the Limestone Wren Babbler, with a subtle wing-bar of small white spots (tips to the greater covert feathers) and a line of white spots along the rear edges of the folded wings (tips to tertial and secondary feathers). It is usually encountered in singles or pairs, and has a varied song consisting of shrill, ringing whistles, and scolds with a hard rattling in alarm.

EYEBROWED WREN BABBLER *Napothera epilepidota* 10–11cm

This very small but rather plump-looking babbler with short wings and a short tail is a fairly common resident of hill and montane evergreen forests in Laos and most of Vietnam (except the far south). Its buffish supercilium contrasting with the dark eye-line, and two rows of white spots on each wing (tips to the greater and median covert feathers), distinguish it from the Streaked and Pygmy Wren Babblers, *N. brevicaudata* and *Pnoepyga pusilla*. The bird's song is a thin, rather drawn-out whistle, *'peeeoo'*, falling in tone. When disturbed or alarmed it usually calls repeatedly with a dry, rattling *'trrrrt'*. Often inquisitive, it is always found on or very near to the ground, and has a habit of hopping along logs, fallen branches or boulders, and flitting through undergrowth.

János Oláh

INDOCHINESE WREN BABBLER *Rimator danjoui* 20cm

James Eaton/Birdtour Asia

This is an enigmatic oddity, previously named the Short-tailed Scimitar Babbler, which is endemic to Vietnam and a strip of central Laos close to the Vietnamese border. It is montane in the southern part of its range, and mainly lowland in the northern part (ascending locally to 1,600m). It has indistinct pale shaft streaks above, a white throat and ventral line, heavy dark streaking across the breast and rufous patches on the neck sides, as well as a distinctive structure, with a longish, slightly decurved bill, shortish tail and rather plump gait. The juvenile is similar to the adult in appearance. Found singly or in pairs, this species is skulking and generally stays close to the ground. It sings with a series of clean, high-pitched and monotone whistles of a little over half a second duration.

COLLARED BABBLER *Gampsorhynchus torquatus* 23–26cm

This distinctive babbler, until recently considered conspecific with the White-hooded Babbler, *G. rufulus*, is a locally common resident except in the far south, where it is absent. Its white hood, pale bill, rufous-brown upperparts, warm orangey-buff underparts and long, white-tipped tail are unmistakable. The race *G. t. luciae* of northern Indochina has a narrow blackish neck collar. The Collared Babbler communicates with a loud, harsh chattering. It often accompanies groups of Red-billed Scimitar Babblers, *Pomatorhinus ochraceiceps*, as well as other babblers and medium-large laughingthrushes in bird waves.

Dave Farrow

VIETNAMESE CUTIA *Cutia legalleni* 17–19cm

Craig Robson

This plump, short-tailed canopy-dweller is a scarce and very local endemic of montane evergreen forests in southern Vietnam and southeast Laos and adjacent central Vietnam. Until recently this species was lumped with the very similar Himalayan Cutia, *C. nipalensis* found in central-northern Laos and northern Vietnam. Both species forage methodically along boughs and trunks covered with mosses and epiphytes. They usually occur in small parties, and sometimes join bird waves. They are sexually dimorphic: the male Himalayan Cutia has the black underpart barring confined to the flank and breast sides. The female has a dark-streaked, grey-brown mantle and back, the Vietnamese Cutia having a dark brown crown and ear coverts, the Himalayan Cutia a dark brown mask and grey crown. The Vietnamese Cutia sings with a variety of strident musical whistles.

ORANGE-BREASTED LAUGHINGTHRUSH
Garrulax annamensis 24–25cm

James Eaton/Birdtour Asia

Only found on the Da Lat and Di Linh plateaux in southern Vietnam, this endemic laughingthrush favours dense patches of vegetation in evergreen forest, secondary growth, forest edge and clearings at between 900 and 1,500m. It has a mostly plain brownish body plumage, but striking head and breast colouration. It has a rich and varied, melodious song, is very skulking and is usually found in pairs; Deo Nui San on the Da Lat Plateau is a good place to find it. Its close relative, the Spot-breasted Laughingthrush, *G. merulina*, lacks the orange colouration and blue facial skin, and has blackish spots on a whitish breast and a buffish-white supercilium behind the eye; it is a local and uncommon resident of northern Laos and Vietnam.

WHITE-CHEEKED LAUGHINGTHRUSH
Garrulax vassali 27–28cm

Janos Olah

This species is endemic to eastern Cambodia, central-southern Vietnam and southern Laos, where it is a locally common resident of degraded evergreen forest, scrub, grass and swidden agriculture at mid-elevations (between 650 and 1,900m). Its plumage is similar to that of the Black-throated Laughingthrush, *Garrulax chinensis*, but the White-cheeked has much brighter rufescent upperparts, bold white tail tips bordered by a blackish subterminal tail band, a less extensive black bib and buffish-brown underparts. The White-cheeked Laughingthrush is especially active and fast-moving, even by laughingthrush standards, and is typically found in flocks that tend to keep to low vegetation. It is not as vocal as many laughingthrush species, however.

BLACK-HOODED LAUGHINGTHRUSH
Garrulax milleti 28–30cm

CAMBODIAN LAUGHINGTHRUSH *G. ferrarius* 28–30cm

Black-hooded Laughingthrush

Cambodian Laughingthrush

Both these species are restricted-range laughingthrushes endemic to the region. The Black-hooded occurs in south-east Laos and central-southern Vietnam (for example on the Da Lat Plateau), where it is a locally common resident of evergreen forest and secondary growth at mid-elevations (800–1,650m), and the Cambodian is confined to evergreen forests in the Cardamom range in south-west Cambodia (Phnom Aural is a good place to find it). The Cambodian has the white restricted to the neck sides, while its hood and breast are browner, and its body plumage is darker grey. Both species travel in noisy flocks, frequenting undergrowth and the lower storeys of forest. They communicate with sudden very loud outbursts of maniacal laughter, rattling and chattering, similar to the vocalizations of the much more common and widespread White-crested Laughingthrush, *Garrulax leucolophus*.

GREY LAUGHINGTHRUSH *Garrulax maesi* 28–30 cm
RUFOUS-CHEEKED LAUGHINGTHRUSH
G. castanotis 28–30cm

Grey Laughingthrush

Rufous-cheeked Laughingthrush

This is a near-endemic species-pair, with a range extending marginally into southern China, and found in evergreen forests at mid-elevations (800–1,600m). The Grey occurs in northern Vietnam (north of Hanoi), the Rufous-cheeked effectively replacing it to the south-west, in the Annamite Mountains of central-northern Laos and Vietnam, with an isolated population on Mount Ba Vi in north-east Vietnam. The species' plumage is predominantly dark grey; the Grey has white ear-coverts, and the Rufous-cheeked has more black on the throat. Both species are rather shy, occur in flocks and vocalize with bursts of noisy laughter, chattering and rattling. The Black-throated Laughingthrush, *Garrulax chinensis*, which is widespread in forest and edge habitats in lowlands and hills throughout the region, looks similar, but has a black throat and breast and a much browner-toned body; it is a fine songster (and a popular cage bird), and is usually found singly or in pairs.

SILVER-EARED MESIA *Leiothrix argentauris* 17–18cm

Martin Hale

This stunningly patterned babbler is a widespread and common resident of hill and montane evergreen forest, secondary growth and scrub, except in the far south and south-west. Males differ from females in being more intensely coloured, and having red uppertail coverts. The bird's dazzling colours make it a target for the cage-bird trade, but it seems to be fairly resilient to trapping pressure in most areas, and can still be found commonly, even in places in close proximity to habitation. It tends to keep to denser vegetation, usually in small parties that are often mixed with other species, and has a loud, cheerful song consisting of a series of 4–5 descending, wavy and musical whistles. It calls with harsh chattering and buzzing notes.

GREY-CROWNED CROCIAS *Crocias langbianis* 22cm

Craig Robson

Probably the rarest of the region's endemic small birds, this sluggish arboreal babbler is restricted to the Da Lat Plateau in southern Vietnam, where it is known from just a handful of sites, including Ho Tuyem Lam, Cam Ly and Chu Yang Sin Nature Reserve. It is Endangered due to loss and degradation of evergreen forest for charcoal production and small-scale agriculture in the narrow altitudinal band (900–1,450m) it inhabits. Its plumage resembles that of the Vietnamese Cutia, *Cutia legalleni*, but it is slimmer and longer tailed, and has streaked, not barred flanks. The Grey-crowned Crocias is generally encountered in singles, pairs and occasionally small groups, joins bird waves and forages slowly for invertebrates, generally keeping to the denser mid-canopy of broadleaved evergreen trees. Listen for its loud and strident 'wip'ip'ip-wiu-wiu-wiu-wiu-wiu-wiu-wiu-wiu' song.

BLUE-WINGED SIVA *Minla cyanouroptera* 14–15cm

James Eaton/Birdtour Asia

A common resident of montane evergreen forest and secondary growth throughout the region, except in far southern Vietnam, this arboreal babbler is structurally fairly slim and long-tailed. It has violet-blue fringing to its primary feathers, forming a panel in the closed wing, and violet-blue fringed tail feathers. The subspecies *M. c. orientalis* of the Dal Lat region of southern Vietnam is larger and more uniformly greyish across the upperparts. The bird forages in small parties, often joining bird waves and favouring the mid-canopy, and when viewed from below it can be recognized by its longish, largely white undertail with black outer border.

SPECTACLED BARWING *Actinodura ramsayi* 24cm

Distinctive and long-tailed, this babbler is a common resident of montane regions in northern and central Laos and northern Vietnam. It inhabits a variety of forested and degraded habitats, including scrub, grass and clearings. It has bold white spectacles, a slight crest, narrow black bars across its wings and a long and narrowly white-tipped tail (the latter feature only obvious from below). It is usually encountered singly or in pairs, sometimes in small groups mixed with other babblers foraging in the lower-middle storeys, and thicker vegetation outside forest. Its song is a mournful-sounding, rather rapid series of undulating whistles, dropping in pitch, and it calls with various harsh notes.

Dave Farrow

BLACK-CROWNED BARWING *Actinodura sodangorum* 24cm

János Oláh

Closely related to the Spectacled Barwing, *A. ramsayi*, this bird is endemic to a small area of south-east Laos and west-central Vietnam (the Kontum/Ngoc Linh region), where it is locally common in evergreen forest, forest edge, scrub and grass, including the understorey of open pine woodland, on plateaux and mountainsides at between 1,100 and 2,400m. Its song and calls are similar to those of the Spectacled Barwing, but its plumage shows several differences, including a black central crown, mostly black wings with a buff patch on the shoulder, and blackish throat streaks. One of the most accessible sites to find this bird is the Lo Xo Pass, where the birds shown here were photographed.

WHITE-BROWED SHRIKE BABBLER
Pteruthius flaviscapis 16–17cm

This is a compact, canopy-dwelling 'babbler' that is one of four shrike babblers in the region, which may in fact be most closely related to the vireos of the New World. This species is a relatively common and widespread resident of montane evergreen forest throughout the region. The male (depicted) has a black head with a broad white supercilium, pale grey upperparts and black wings with flame- to chestnut-coloured tertials, varying with the race. The female also has the white supercilium, less contrasting against a pale grey head, and has a bright yellowish-green wing patch and tail. The bird's strident song is a rhythmic series of 3–5 clear notes, repeated many times. It forages along large canopy and mid-canopy boughs.

János Oláh

INDOCHINESE BUSHLARK *Mirafra marionae* 14–15cm

One of three larks in the region, this species is fairly widespread and locally common in relatively dry habitats with scattered trees, and deciduous dipterocarp woodland of the plains and lower hills, in Laos, Cambodia and central-southern Vietnam. It is distinguished from the Australasian Bushlark, *M. javanica*, which inhabits short grassland and dry paddies, by its bolder breast streaking and lack of white outer tail feathers, and from the Oriental Skylark, *Alauda gulgula*, which occurs in various open habitats, by its lack of a prominent crest, stouter bill and more rufous wings. The Indochinese Bushlark gives its sibilant, see-sawing song from a perch or short, low song-flight. Both the Oriental Skylark and

James Eaton/Birdtour Asia

Australasian Bushlark have towering song-flights; that of the former is almost incessant and high-pitched, while that of the latter comprises short phrases with mimicry, also delivered from perches.

SCARLET-BACKED FLOWERPECKER *Dicaeum cruentatum* 9cm

Flowerpeckers are a group of tiny, canopy-dwelling birds. The Scarlet-backed Flowerpecker is the region's most widespread and common, found in open forest, forest edge, cultivated areas, parks and gardens in lowlands and hills, and is one of the few bird species occurring in the heart of larger cities and towns. The male's plumage is striking (the bright red crown extends as a broad stripe right down his back to his rump), but the female is more subtle – the red is restricted to her rump and uppertail coverts, the remainder of her upperparts are dull olive-grey and her underparts are greyish-white. The bird's call, which is similar to the calls of some other flowerpeckers (such as the Yellow-vented Flowerpecker, *D. chrysorrheum*), is a hard and metallic *'dick'* – it is a very useful aid to locating it as it forages in the crowns of flowering and fruiting trees.

Dave Farrow

GREEN-TAILED SUNBIRD *Aethopyga nipalensis* 11–13cm

Male (above; female (right)

This colourful sunbird is a locally common resident of montane areas in Laos and parts of north-west, central and southern Vietnam, inhabiting montane evergreen forest, forest edge and secondary growth. The male could be mistaken for the more widespread Black-throated Sunbird, *A. saturata*, which occupies similar habitats at lower elevations, but that species has a black upper breast and much paler underparts. The female is very similar to both Mrs Gould's and Black-throated Sunbirds, *A. gouldiae* and *A. saturata*, but lacks a well-defined pale rump band and has obvious white tips to the underside of her graduated tail. Sunbirds are mainly nectivorous, and perform similar ecological functions to the hummingbirds in the Americas.

FORK-TAILED SUNBIRD *Aethopyga christinae* 10–12cm

This is a near endemic whose range extends into southern China. It is a fairly common resident of evergreen forest and secondary growth in lowlands and hills (at up to about 1,400m) through much of Vietnam and adjacent eastern parts of central-southern Laos. The male (depicted) has well-separated, short extensions to its central tail feathers, giving rise to the species' name. The female is like the female Crimson Sunbird, *A. siparaja*, but has prominent white tips to the undersides of the tail feathers and is slightly more yellow below. The bird's call is a slightly explosive, high-pitched *'swit'*. Like other sunbirds, it has a brush-tipped, tubular tongue adapted to feeding on nectar.

128

CRIMSON SUNBIRD *Aethopyga siparaja* 11–13cm

Rob Tizard

Another of the longer-tailed sunbirds, the Crimson Sunbird is a widespread and common resident of lowland and hill evergreen and mixed deciduous forests, secondary growth and even gardens. The male is unmistakable, although the male Mrs Gould's Sunbird, *A. gouldiae*, is similarly coloured and patterned, but occurs at higher elevations. The female has the distinction of being the dullest of the sunbirds, and can be identified by her uniform dull olive colour, the lack of any white in her tail, her pale rump band and (typically) the lack of yellowish colour on her underparts. Young and eclipse males (undergoing a post-breeding body moult) look like the female but have a red wash across the breast.

LITTLE SPIDERHUNTER *Arachnothera longirostra* 16cm

James Eaton/Birdtour Asia

Larger and more robust than the sunbirds, this spiderhunter is a common and widespread resident throughout the region, favouring evergreen forest, secondary growth, cultivated areas and gardens from the plains up to mid-elevations. In Indochina this spiderhunter is only really confusable with female sunbirds, from which it is distinguished by its larger size, longer and stouter bill, whitish throat, broken eye-ring and dark moustachial streak. It frequents the understorey, particularly favouring the broad fronds of banana plants (both wild and cultivated), and can be located by its loud and sharp '*chit*' call, which is frequently repeated. It sings with a rapidly repeated '*wit-wit-wit-wit-wit-wit...*'

STREAKED SPIDERHUNTER *Arachnothera magna* 17–20cm

Craig Robson

The larger of the region's two spiderhunters, this is also a common and widespread resident, although it is absent from most of Cambodia except the north-east. The Streaked Spiderhunter favours similar habitats to the Little Spiderhunter, *A. longirostra*, but tends to be scarcer in lowlands and more prevalent in hillier terrain, its range extending up to 1,800m or so. The combination of liberal blackish streaking over bright yellowish-olive upperparts and white underparts, and orange feet, is unmistakable. The bird has a strident, chattering song, and calls with a loud '*chittik*', often in flight, which is strongly undulating or bounding (recalling some woodpeckers). As their name implies, spiderhunters forage on spiders, although they have long bills and tongues well adapted to feeding on nectar.

PLAIN-BACKED SPARROW *Passer flaveolus* 14–15cm

This South-East Asian endemic, the most brightly coloured sparrow in the region, is a locally common resident of Laos, Cambodia and central-southern Vietnam, where it occurs in relatively open habitats with scattered trees, cultivated areas, coastal scrub and villages, up to mid-elevations. The male's (depicted) plumage of chestnut, yellow, grey and black is unmistakable, but the female is more subtle. She can be told from other sparrows by her unstreaked upperparts and yellowish-tinged underparts. This species occurs in pairs and small flocks, often with other sparrows, and frequently nests and roosts colonially.

Dave Farrow

EURASIAN TREE SPARROW *Passer montanus* 14cm

The most abundant sparrow in the reg-ion, this species is found throughout and is almost ubiquitous around human habitation, from inner cities to montane villages (at up to 1,800m). The plumage of the sexes is similar. It could be confused with that of male House and Russet Sparrows, *P. domesticus* and *P. rutilans*. The House Sparrow has recently expanded into the region's towns and agricultural areas (for instance into much of the deforested Mekong plain). It has a grey cap, no black ear spot and a pale bill. The Russet Sparrow

Rob Tizard

is found in montane swidden agriculture and forest edge in the far north; it has a wholly chestnut crown, upperparts with black scales on the mantle and buff (not white) cheeks with no black ear spots. Both the Eurasian Tree Sparrow and the House Sparrow nest and roost communally, often in noisy gatherings of many hundreds or thousands.

FOREST WAGTAIL *Dendronanthus indicus* 17–18cm

Martin Hale

A rather atypical wagtail, this is an uncommon to fairly common passage migrant and winter visitor to the region from its breeding grounds in central-northern China, Korea, Japan and Siberia. It is found in a variety of habitats, from forest to cultivated areas with trees, gardens and mangroves, at low to mid-elevations. By day it is usually encountered on its own, often walking slowly and deliberately on leaf litter well away from water. In some areas it roosts communally, sometimes large numbers. Its boldly patterned plumage is very distinctive, consisting of a double black breast band, black and creamy bars in the wing, pale creamy supercilium and dull olive upperparts. It calls with a metallic 'pink' or 'zink', especially in flight.

MEKONG WAGTAIL *Motacilla samveasnae* 19cm

Peter Davidson

Endemic to the lower Mekong Basin, this recently described resident wagtail occurs along the Mekong and its larger tributaries in southern Laos, northern Cambodia, south to Kratie, and Yok Don National Park in Vietnam. It can be distinguished from the black-backed forms of the White Wagtail, *M. alba*, by its white supercilia, black forecrown and the isolated white spot on each neck side. It breeds on vegetated islands, and during the flood season (July–November) feeds along exposed sandy and muddy upper banks and the adjacent floodplain. It is Near Threatened because the many proposed Mekong Basin dams will potentially flood long stretches of its riverine range. Effective conservation of this and other riverine wetland species requires international cooperation. Kampi Pool just north of Kratie is an excellent spot to find it during the dry season (December–May).

WHITE WAGTAIL *Motacilla alba* 19cm

Martin Hale

Highly variable, this wagtail is represented by several subspecies in Indochina, which can be grouped into grey- and black-backed forms, sometimes treated as separate species. It is both a common winter visitor and passage migrant (especially in southern and central Indochina), and an uncommon resident in the far north, found around all manner of wetland habitats, from puddles to large river shores; sandbars in wide rivers are particularly favoured. All races have a black and white head pattern and black breast gorget, and either black or pale grey upperparts with black and white wings, and a black tail with white outer feathers. The bird calls with a clear, strident, disyllabic 'chissick'.

PADDYFIELD PIPIT *Anthus rufulus* 17cm

This is a common and wide-spread resident of open and relatively dry, short-grass habitats, in particular areas of cultivation and playing fields, and sparsely vegetated ground, from lowland plains up to middle elevation plateaux. It looks very like Richard's Pipit, *A. richardi*, which is a common winter visitor and passage migrant to Indochina from China, Mongolia and Siberia, but it is smaller, has a relatively shorter tail and is usually less strongly streaked on the breast. The best way to distinguish these two species is by their

János Oláh

calls: the Paddyfield gives an abrupt '*chup*', Richard's a harsher, more explosive '*shreeep*'. The Paddyfield has an undulating song flight, as it rises high into the air and parachutes down.

OLIVE-BACKED PIPIT *Anthus hodgsoni* 16–17cm

Martin Hale

This Palearctic breeder is a winter visitor to Indochina and the most common pipit of wooded habitats, including pine and deciduous dipterocarp woodland, evergreen forest tracks and edges, and cultivated areas, occurring at all altitudes, and even in parks and gardens on passage. It is the only pipit with rather plain olive upperparts; other useful identification features include the prominent whitish supercilium and pale creamy spot on the ear coverts, and the bold black spotting on the breast and flanks. The bird can be rather secretive, and has the distinctive habit of gently but persistently pumping its tail. It usually flies up to perch in trees when flushed, and calls with a thin but rather hoarse '*tzzzz*' or '*teez*'.

RED-THROATED PIPIT *Anthus cervinus* 15–16cm

Martin Hale

This pipit is locally common as a winter visitor to the region from its breeding grounds in far northern Russia, favouring open places, especially paddies and other cultivated areas in proximity to water. Adults have a brick-pink wash to their head and upper breast, which is usually less pronounced on females and in non-breeding plumage. Immature birds lack the pink colouration, but are distinguishable from other pipits by their bold black moustachial stripe and underpart streaking, buffish-white mantle braces and streaked rump. The species usually occurs in flocks, often close to livestock, which disturbs insects on which the bird feeds, and many thousands of individuals may roost together, for example in the Xe Pian National Protected Area, in southern Laos, and on the Tonle Sap floodplain. The bird has a distinctive high-pitched but strident 'psseeoo' or 'pseee' call note.

ASIAN GOLDEN WEAVER *Ploceus hypoxanthus* 15cm

Allan Michaud

This is the rarest of the region's three species of weaver, and it is considered Near Threatened due to conversion of its wetland habitats for agriculture, and to trapping for the cage-bird and merit-release trades. It inhabits lowland marshes, grassland, reeds, sedges and traditional ricefields, mostly close to water. The plumage of the breeding male is unmistakable and stunning. In female and non-breeding plumages, the bird can be distinguished from the Baya Weaver, *P. phillipinus*, by its massive, deep-based bill (it is the largest billed of the three weavers in Indochina), plain breast lacking any hint of mottling and stronger crown streaking. It nests colonially during the wet season (April–October), building a ball-shaped nest with a side entrance in reeds, tall grass or large waterside bushes.

SCALY-BREASTED MUNIA *Lonchura punctulata* 12cm

This small, compact and gregarious, finch-like bird is a common and widespread resident of lower and mid-elevations throughout the region. It is a specialist seedeater, and makes local seasonal migrations in order to exploit different food sources, enabling it to breed all the year round. It thrives in degraded habitats, mostly cultivated areas, scrub, secondary growth and even urban gardens. The adults are distinctive with their warm brown ground colour and pale-scaled underparts, while the juveniles are nondescript drab buffish. This bird is a fast, direct flier travelling in tight flocks, and usually calling with a thin, piping '*ki-dee, ki-dee...*' Its flocking habit makes it straightforward to trap, and many birds are caught for the cage-bird trade and merit release (for example along the Mekong riverfront in Phnom Penh).

Allan Michaud

VIETNAMESE GREENFINCH *Carduelis monguilloti* 14cm

Endemic to the Da Lat Plateau in southern Vietnam, where it is locally common in open pine forest at mid-high elevations (1,050–1,900m), this species is globally Near Threatened, partly because its tiny range makes it very susceptible to extinction. It may in fact, however, benefit from the clearance of evergreen forest that threatens several of the other Da Lat region's endemics: the growth of its favoured pine *Pinus kesiya* forest habitat is stimulated by fire clearance and is increasing, although the potential for rapid clearance remains. The Vietnamese Greenfinch is the only greenfinch found in the Da Lat area, and as such is unmistakeable. The Grey-capped Greenfinch, *C. sinica*, occurs locally along the nearby

János Oláh

coast, and the similar looking Black-headed Greenfinch, *C. ambigua*, is a local resident in the north of the region (for instance around Sapa in Vietnam and Xiangkhouang in Laos).

RED CROSSBILL *Loxia curvirostris* 17cm

Craig Robson

An outlying population of this chunky finch, with its unique bill adapted for extracting seeds from pine cones, occurs on the Da Lat Plateau and, with the nearest population of the Red Crossbill occurring in the Himalayas, it is variously treated either as a subspecies, *L. c. meridionalis*, or as a separate species. Like the Vietnamese Greenfinch, *Carduelis monguilloti*, the Red Crossbill inhabits open pine forest at mid-high elevations; it is therefore faring better than the evergreen-forest endemics on the plateau. The male has largely orange-red plumage with dark brownish wings and tail, while the female (depicted) is dull brownish-grey with green flecking and darker streaking. The bird is usually found in small flocks, and in flight calls with a repeated, hard and rather loud *'jip'*.

SPOT-WINGED GROSBEAK *Mycerobas melanozanthos* 23cm

Dave Farrow

This massive-billed, large-headed finch is an uncommon resident of montane forest, secondary growth and forest edge in northern Laos and north-west Vietnam. Both sexes have a huge pale grey, conical bill, a white patch in the primaries most visible in flight, and white or pale yellowish spots forming a bar across the closed wing and bold tips to the tertial feathers. The male is otherwise black with bright yellow underparts, the female (depicted) paler yellow with blackish streaks and a boldly striped head pattern. The birds are usually found in flocks, which are often fairly large, and rove for trees in fruit, their primary food source. They call with a rattling trill, flocks making a cackling chorus, and have two- or three-note melodious and mellow song phrases.

YELLOW-BREASTED BUNTING *Emberiza aureola* 15cm

Martin Hale

Formerly abundant and widespread as a winter visitor from northern Japan, China, Mongolia and Russia, this bird is now only a locally numerous winter visitor. It was recently classified as globally Near Threatened because of severe declines in some breeding areas (perhaps drought related) and concerns over trapping at migration and wintering sites. It frequents dry paddies and other cultivated areas, grassland and scrub. Its size and proportions are sparrow-like, and in winter plumage (depicted) it looks not dissimilar to a sparrow, with a variable yellowish wash to the underparts, dark streaking on the flanks, white outer tail feathers and, on the male, a white shoulder patch and dark ear covert mask. It is often found in loose flocks, which are sometimes very large (especially at roost in tall grass or reeds), and calls with a short, metallic '*tsip*', rather like several other buntings.

CHESTNUT BUNTING *Emberiza rutila* 14cm

James Eaton/Birdtour Asia

Another boreal breeder, this species is a locally common winter visitor to forest understorey, forest edge, swidden agriculture, grass and scrub, mostly in the foothills and mountains of Laos and Vietnam. As in all buntings, the sexes differ. The male is chestnut above and yellow below (more subdued in winter than when in breeding plumage, as depicted, acquired in March–May), while the female looks rather like a female Yellow-breasted Bunting, *E. aureola*, but has a less contrasting head pattern, as well as a rufous-chestnut rump and uppertail coverts. The Chestnut Bunting likes feeding on seeding bamboo, so search for it where bamboo has recently flowered. It occurs solitarily or in small groups, and usually flies up into trees when flushed.

GLOSSARY

Bird wave A flock of birds travelling and feeding together.

Brood parasites Certain birds that lay their eggs in the nests of other birds and do not provide any parental care for their offspring.

Cap Well-defined patch of colour or bare skin on top of the head.

Cere The soft skin at the base of a bird's bill.

Crepuscular Most active at dawn and dusk.

Culmen Ridge of the upper mandible between the nasal holes and the tip.

Diurnal Active during the day (in reference to typically nocturnal birds).

Endemic Restricted to a defined area, usually small on a global scale.

Eye-ring Area immediately surrounding the eye.

Eye-stripe A concolourous line running through the eye.

Flight feathers The primary, secondary, tertial and tail feathers.

Genus A group of species more closely related than a family.

Hand (of wing) The spread primary feathers (typically of raptors) in flight.

Jowl The appearance created by a heavy, rounded cheek and throat, for example in the Cattle Egret.

Kleptoparasitism Form of feeding where one animal takes prey from another that has caught, killed or otherwise prepared it.

Lore In birds, the surface on each side of the head between the eye and the upper base of the beak or bill.

Mandible The upper and lower mandibles form a bird's beak.

Mantle The upper back feathers.

Mask A variable dark plumage patch that covers the eye.

Mesial Of the middle (usually refers to a central throat stripe).

Montane Of or inhabiting mountainous country.

Moult The shedding of old feathers and growth of new ones.

Moustachial stripe A line bordering the throat at 45 degrees to the bill.

Nuchal collar Band across the back of the neck.

Passerine Belonging to the avian order Passeriformes, which includes the perching birds. Passerine birds make up more than half of all living birds.

Primary feathers Flight feathers of the outer wing joint.

Sallies/sallying Sudden flight from perch to catch prey or take fruit.

Secondary feathers Flight feathers of the inner portion of the wing.

Sexually dimorphic Male and female have different plumages.

Shawl A uniform patch of colour on the nape, head sides and mantle.

Sub-terminal band Just short of the tip or end (for example of the tail).

Supercilium The eyebrow, or line above the eye.

Tail coverts Feathers covering the bases of the tail feathers.

Terminal band At the tip or the end (for example of the tail).

Tertial feathers The innermost flight feathers, borne at the base of the wing.

Wing lining The inner part of the underwing, often the underwing coverts.

Wing coverts Feathers covering the primary and secondary feather bases.

Worn (plumage) When the feathers are old and abraded before moulting.

FURTHER READING

BirdLife International, 2003, *Saving Asia's Threatened Birds: a Guide for Government and Civil Society*. An overview of the conservation issues facing the key habitats for Asia's threatened birds.

Collar, N.J., A.V. Andreev, S. Chan, M.J. Crosby, S. Subramanya, and J.A. Tobias, 2001, *Threatened Birds of Asia CD-ROM*, Cambridge: BirdLife International. Incredibly detailed work summarizing all known distribution, status, ecological and threats information on threatened birds and related policy recommendations in the region.

Del Hoyo, J., A. Elliott and D.A. Christie, eds, 1992–2007, *Handbook of the Birds of the World*, Volumes 1–12, Barcelona: Lynx Edicions. www.hbw.com. The most comprehensive overview of the world's birds published to date, including a wealth of useful information, colour plates and photographs to complement this guide.

Duckworth, J.W., R.E. Salter and K. Khounboline, 1999, *Wildlife in Lao PDR: 1999 status report*, Vientiane: IUCN/WCS/CPAWM. www.wcs.org/international/Asia/Lao PDR/wildlifeinlaopdr. Detailed status, distribution and conservation information for the fauna of Laos, including all bird species.

Lekagul, B. and P.D. Round, 1991, *A Guide to the Birds of Thailand*, Bangkok: Saha Karn Bhaet Co. Ltd. Groundbreaking identification guide covering many species that also occur in Cambodia, Laos and Vietnam.

Robson, C.R., 2000, *A Field Guide to the Birds of South-East Asia*, London: New Holland. The most comprehensive and up-to-date field guide to the region's birds: indispensible. Second edition 2009.

Robson, C., 2005, *New Holland Field Guide to the Birds of South-East Asia*, London: New Holland. A concise paperback version of Robson (2000).

Thewlis, R. M., R. J. Timmins, T.D. Evans and J.W. Duckworth, 1998. 'The conservation status of birds in Laos: a review of key species', *Bird Conservation International* 8 (supplement): 1–159. Much more site-specific detail than Duckworth *et al.* (1999) for species of conservation concern.

WEB RESOURCES

Important Bird Areas of Cambodia, Lao PDR and Vietnam. This is available for download from www.birdlifeindochina.org. It contains detailed bird status, habitat and conservation data on the most important areas for birds of conservation concern.

The Oriental Bird Club, www.orientalbirdclub.org, is a must for anyone who is planning a trip to the region. It produces a biannual birding and conservation bulletin, *Birding Asia*, as well as an annual journal, *Forktail*. 'OBC' also supports an extensive Web database containing bird images. It can be visited at http://orientalbirdimages.org.

ACKNOWLEDGEMENTS

Many people contributed to this book. I am extremely grateful to each of the photographers, James Eaton, Dave Farrow, Martin Hale, Rob Hutchison, Allan Michaud, Pete Morris, János Oláh, Bill Robichaud, Craig Robson, Rob Timmins, and Rob Tizard, for providing their superb images to illustrate this guide. Furthermore, their generosity has enabled close to US$3,000 to be donated to conservation in the region. Craig Robson assisted with numerous technical queries. Will Duckworth commented extensively on a draft of the text, and Colin Poole also suggested improvements. The Wildlife Conservation Society (WCS), in partnership with the Ministry of Agriculture and Forestry in Laos, and the Ministry of Agriculture, Forestry and Fisheries, and Ministry of Environment in Cambodia, supported most of my work in those two countries: it was my privilege to work and share experiences with the staff of WCS and its partners. Krystyna Mayer and Simon Papps at New Holland commissioned and guided the book through to publication. Conversations and shared field experiences with Paul Batchelor, Tom Clements, Will Duckworth, Jonathan Eames, Tom Evans, Dave Farrow, Frédéric Goes, Hong Chamnan, Le Manh Hung, John Parr, Colin Poole, Craig Robson, Bill Robichaud, Phil Round, Tan Setha, Chris Shepherd, Dave Showler, Rob Timmins, Rob Tizard, Joe Tobias, Nguyen Duc Tu, Chanthavi Vong-khamheng, Joe Walston and James Wosltencroft fostered my knowledge of the region's birds. I heartily thank you all. The people of Laos, Cambodia and Vietnam treated me to their warm hospitality during the eight or so years I lived and worked in the region. Finally, thanks to my wife Julie and son Bram for their encouragement and understanding throughout.

INDEX

KEY TO COLOURED TABS

Cormorants, anhingas & frigatebirds

Pelicans, herons, storks, cranes, ibises & spoonbills

Ducks

Raptors

Gamebirds

Rails & bustards

Waders

Gulls & terns

Pigeons, doves & parrots

Owls & nightjars

Kingfishers. bee-eaters, rollers & hoopoes

Swifts, swallows & martins

Hornbills, trogons & cuckoos

Barbets & woodpeckers

Broadbills & pittas

Wagtails, pipits & larks

Bulbuls

Minivets & leafbirds

Flycatchers, chats & niltavas

Thrushes

Warblers, tailorbirds & parrotbills

Babblers & relatives

Tits

Treecreepers & nuthatches

Sunbirds, spiderhunters & flowerpeckers

Orioles & drongos

Shrikes, woodswallows & cuckooshrikes

Crows

Starlings & mynas

Sparrows, finches & buntings

144